DRINK MAPS
IN VICTORIAN BRITAIN

DRINK MAPS
IN VICTORIAN BRITAIN

KRIS BUTLER

BODLEIAN
LIBRARY
PUBLISHING

I raise a glass of English bitter pulled from a cask
in dedication to my mother and late father, Cindy and Harvey Butler.

And another glass – a hoppy beer – to my late friend Regine Gerhardt,
who was there when I first saw a drink map and said it made her thirsty.

First published in 2024 by Bodleian Library Publishing
Broad Street, Oxford OX1 3BG
www.bodleianshop.co.uk

ISBN 978 1 85124 578 9

Publisher: Samuel Fanous
Managing Editor: Susie Foster
Editor: Janet Phillips
Picture Editor: Leanda Shrimpton
Cover design by Dot Little at the Bodleian Library
Designed and typeset by Lucy Morton of illuminati in 10½ on 16 Baskerville
Printed and bound in China by 1010 Printing International Ltd
on 157 gsm Chinese Matt Art paper

British Library Catalogue in Publishing Data
A CIP record of this publication is available from the British Library

CONTENTS

ACKNOWLEDGEMENTS

Beers are raised in gratitude to:

Nick Millea, who first showed me the *Drink Map of Oxford* at a cartography conference in Budapest, Hungary, in 2005. I was instantly smitten, and he has supported my curiosity with key introductions and clever suggestions over many years and beers.

To my first readers for their valuable time, insightful comments and suggestions: Beatriz Ashfield, Keith Ashfield, Cindy Butler, Ann Cortissoz, Ann McClenahan, Nick Millea and Donna Sherman.

I am grateful to the wonderful law firm of Holland & Knight for allowing me to take a leave of absence to finish my research, and especially to Carrie Weintraub and Jaime Kaplan.

I clink pints of real ale with everyone who put me up for a night or a week, said nice things or tough things nicely, brewed beers to go with my map talks, funded travel, pub tabs and scotch eggs, invited me to chat on BBC radio, pointed to resources I would not have known about, assured me it was perfectly normal over the years when I said I was spending yet another vacation searching for obscure maps in the UK and who otherwise shared their time, expertise and humour: Stuart Ackland, Claudia Asch, Peter Barber, Gjis Boink, Tony Campbell, Yvan de Baets, Lena Denis, Paulette Edwards, Bronagh Fay, Chris Fleet, Janie Gottwalt, Ron Grim, Tom Harper, Jim Houglan, Dann and Martha Holley-Paquette, Joanne Limbach, Emily Mann, Will Meyers, Alice Millea, John Moore, Katie Parker, Megan Parisi, Dave Pickersgill, Jon and Marie Raney, Jonathan Rosenwasser, Brian Shurtleff, Chet van Duzer and everyone in the Tines & Steins book club.

Finally, copious toasts of vintage champagne to the Bodleian Library Publishing team and all the librarians, archivists and protectors of paper everywhere. Thank you.

Map of England and Wales showing *Geographical Distribution of Drunkenness*, 1886. From *The Temperance Problem and Social Reform* by Joseph Rowntree and Arthur Sherwell, 1900.

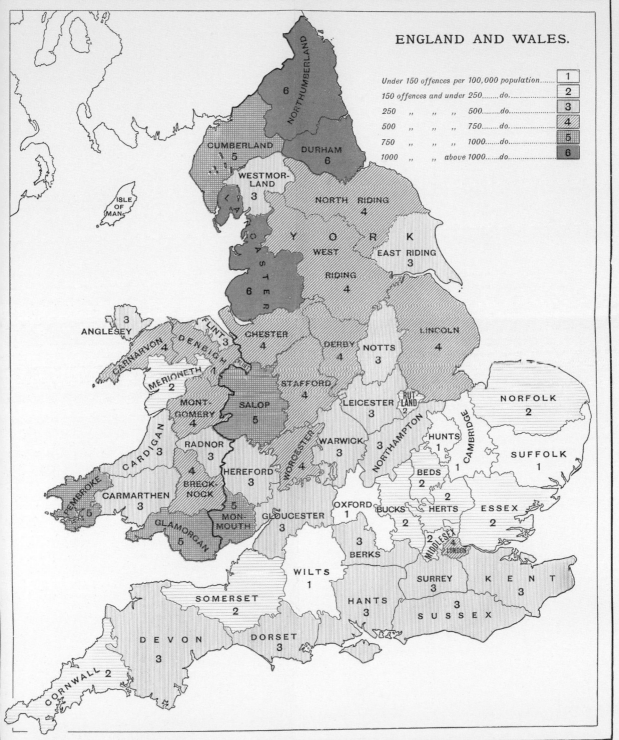

ENGLAND AND WALES.

Under 150 offences per 100,000 population....... 1
150 offences and under 250........do................ 2
250   „   „   „   500........do................ 3
500   „   „   „   750........do................ 4
750   „   „   „ 1000........do................ 5
1000   „   „   above 1000......do................ 6

NORTHUMBERLAND 6

CUMBERLAND 5   DURHAM 6

WESTMOR-LAND 3

ISLE OF MAN

NORTH RIDING 4

LAN-CASTER 6   YORK

WEST RIDING 4   EAST RIDING 3

ANGLESEY 3

FLINT 3

CHESTER 4

LINCOLN 4

CARNARVON 4   DENBIGH

DERBY 4   NOTTS 3

MERIONETH 2

STAFFORD 4

MONT-GOMERY 4   SALOP 5

LEICESTER 3   RUT-LAND 2

NORFOLK 2

CARDIGAN 3

RADNOR 3

WARWICK 3

NORTHAMPTON

HUNTS 1

CAMBRIDGE 1   SUFFOLK 1

HEREFORD 3

WORCESTER 4

BEDS 2

CARMARTHEN 3   BRECK-NOCK 4

MON-MOUTH 5

GLOUCESTER 3

OXFORD 1   BUCKS 2   HERTS 2   ESSEX 2

PEMBROKE 5

GLAMORGAN 5

BERKS 3

MIDDLESEX 2   LONDON 4

WILTS 1

HANTS 3

SURREY 3   KENT 3

SOMERSET 2

DEVON 3

DORSET 3

SUSSEX

CORNWALL 2

Walker & Boutall sc.

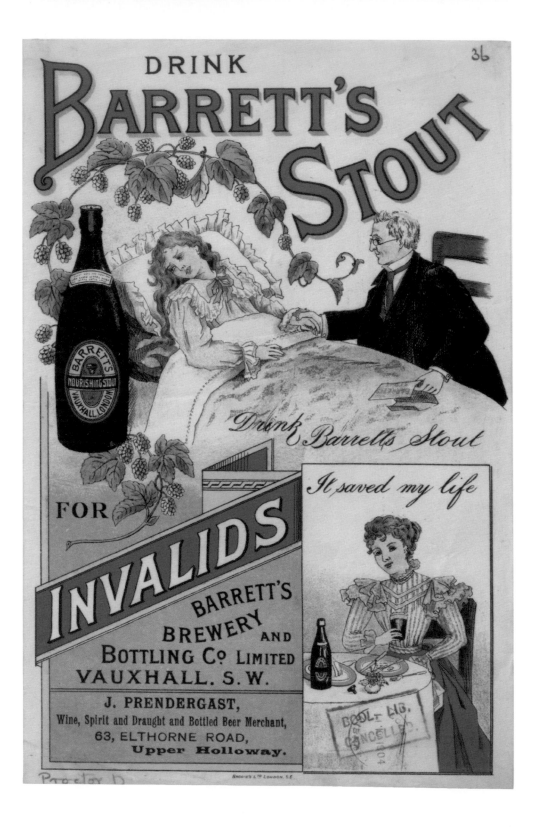

CHAPTER ONE

THE DRINK PROBLEM

This is the story of drink maps, and it's probably not what you think. It's not about pub crawls or plotted ale trails. Instead these are maps with an agenda that was adamantly hostile to drinking alcohol, made by an organized faction known as the Temperance Movement. The logic at the time of the maps' creation went as follows: if people are shown how many places there are to buy alcohol, they will be so appalled that they will join the effort to end drinking. In hindsight this logic is obviously flawed. How could anyone believe it was a good idea to proffer a pub guide to convince onlookers to stop drinking? Yet, as you will see, there are many reasons why this made sense at the time.

WHAT IS A DRINK MAP?

A drink map was a map of a specific place, usually a single town or city, highlighting where alcohol was sold with the intent to convince people to demand that there be fewer of them. Some had statistics printed on them, such as the number of drink-related crimes the previous year or

1 This advertisement is an example of the common perception that beer was healthy. The insert shows a recovered woman upright and drinking beer, thinking 'It saved my life'. Advertisement for Barrett's Stout, c. 1904.

the number of liquor licences (state-regulated permits to sell alcohol) compared to the population. Others had long solemn articles, shocking statistics or colourful graphs. The term 'drink map' is an inclusive label, and people of late-nineteenth-century Britain understood that it meant a map with a temperance goal. A core group of maps from this era had titles that included the words 'drink map' while others, even if they did not have the word 'drink' on them, were referred to as drink maps in contemporaneous newspaper articles, recorded meeting minutes and sworn testimony. Others were neither titled nor referred to as drink maps, yet had that persuasive ambition in mind. In this book the phrase 'drink map' inclusively refers to maps signifying places to buy intoxicants with the intent to reduce their number, including many identified in historical records but no longer extant (or yet to be found – see APPENDIX for all maps identified in the research for this book).

TEMPERANCE

The word 'temperance' broadly means moderation and a consistent, level approach. The root word 'temper' refers to exerting control and exercising restraint. Think of tempering steel or chocolate – a process of gently and precisely heating and cooling until the result is stronger, shinier and more resistant to breaking than before. Temperance is also the name of the social movement of restraint specifically relating to alcohol. At first it was primarily a belief in moderation. Early proponents of moderate drinking formed groups to exert organized social and moral pressure on others to convince them to drink less or at least refrain from hard liquor. These efforts were followed by a division within the movement, where extremists embraced a more dramatic position known as teetotalism which demanded complete abstinence from alcohol. This split between moderation and abstinence has always caused tension within temperance movements everywhere. In the United Kingdom it sometimes thwarted temperance efforts by alienating those who believed it reasonable to consume reduced amounts of alcohol. Even Queen Victoria spoke out against total abstinence, believing it an unreasonable impossibility – though necessary in some individual cases. Just like any group bent on persuading others of their position, defining that position was sometimes its own fight.

HISTORICAL CONTEXT

To fully appreciate why the idea of drawing maps of places to find alcohol ever seemed like an effective way to convince people to drink less of it, it's necessary to understand the background of the time the maps were made, which was primarily the mid- to late 1800s in England, in the Victorian era.

At the start of the century, the vast majority of people could not read or vote, and therefore had little ability to change their situations; they had no voice and were powerless to push back on long hours and dangerous working conditions, or to argue for any kind of safety nets if they were injured. Shortly before Queen Victoria ascended in 1837, the Great Reform Act of 1832 had passed, which redistributed power through voting rights in an attempt to correct imbalances and abuses, and gave more say to the people about how they were governed. Among other things, it broadened the qualification to vote and was an early step in a direction of reforms that would eventually define the century. But even after the Act, the vast majority of working men still did not qualify – fewer than 20 per cent could vote. Nor any women. On top of that, ballots were not secret, and therefore more open to bribery, peer pressure and beer pressure.

Meanwhile simultaneous and compounding innovations had a dramatic impact on daily life. People went from using candles, outhouses and horses to enjoying electricity, indoor plumbing and trains over this century. Reliable and speedy railway operations also meant that goods such as beer and newspapers could get around much faster. As the

TIMELINE OF RELEVANT LAWS
AND OTHER MILESTONES

1828 Alehouse Act consolidates previous licensing laws and establishes annual licensing days.

1830 Beerhouse Act establishes unregulated trade in beer with intent to divert people from spirits; beer sales remained unregulated until 1869.

1837 Victoria ascends to the throne at eighteen years of age.

1838 The first modern steam-powered railway opens.

1850 The first workhouses opens to provide bed and food (and beer) for the poor in exchange for their work.

1851 The 'Maine Law' passes in the US state of Maine, the first to outlaw the sale of alcoholic beverages.

1851 The first free public library opens in Winchester.

1852 The first public flushing toilet opens in London.

1853 The Vaccination Act requires children to be vaccinated against smallpox, and mandates punishments for non-compliant parents.

1853 Formation of the United Kingdom Alliance.

1858 India comes under British rule.

1858 A particularly stinky summer leads to demand for a modern sewer network in London.

1862 Sir Wilfrid Lawson MP first introduces the Permissive Bill Resolution in the House of Commons.

1863 The first underground railway opens in London.

1864 Louis Pasteur applies heat to kill microbes in wine and milk (a method already used in China since the 1100s and Japan from the mid-1400s). His continued experiments significantly extend the shelf life of many foods, including beer.

1869 The Wine and Beerhouse Act (England and Wales) establish magistrate authority over liquor licences and raise rateable qualification of beerhouses that opened unregulated in 1830, closing hundreds immediately.

1870s Electric lighting debuts, leading to replacement of gas lamps on the streets of London; the first individual bulbs for home use arrive in 1879.

1870	Compulsory education in England for five- to ten-year-olds, dramatically increasing literacy.
1872	The Licensing Act declares that a 'bona fide traveller' could be supplied with liquor when public houses were otherwise closed if the person had travelled at least 3 miles. The Act also prohibits the sale of spirits to children under sixteen years of age, among other things.
1872	The Coal Mines Regulation Act prohibits minors from being paid their wages in drink shops.
1875	Henry Butcher perfects a hop training process which increases production and launches a hop harvesting season in Kent like the French *vendange*.
1877	Act professionalizes clerks' (magistrate advisors) status so they are paid and required to be qualified in law.
1881	Closing of public houses on Sundays in Wales.
1882	The Beer Dealers' Retail Licences Amendment Act clarifies that magistrates have the power to refuse to renew an existing licence. Also known as the 'Off License Act of 1882'.
1883	The Payment of Wages Prohibition Act forbids wages from being paid to workers where liquor is sold.
1886	Brewery-owning MPs peaked at fifty, with most coming from London.
1886	The Intoxicating Liquors (Sale to Children) Act prohibits sale of alcohol to children under thirteen for their own consumption (but they can still fetch it for their parents).
1887	The Truck Amendment Act prohibits payment of agricultural workers in beer or cider. (Up to this time, orchard owners would send trucks out into the fields with refreshments for workers in the form of cider, which would then be deducted from their wages.)
1891	Free education for children aged five to thirteen.
1896	Royal Commission on Licensing Laws appointed. Split report published in 1899.
1898	The Inebriates Act to promote Inebriate Reformatories is passed.
1901	Queen Victoria dies.

populations shifted mid-century from primarily rural to majority urban, drinking habits also changed, moving from a private home activity to a public one (see TIMELINE).

Drinking and trouble have been companions throughout history. Overindulgence in alcohol has spiked at different times, and the mid- to late 1800s in Britain, Europe and North America was one of those moments. Historians point to multiple and often exacerbating causes for this, including a side effect of the Industrial Revolution's newly overcrowded urban areas and condensed work sites like docks, factories and railways; few places to gather for recreation (the local public house was typically used as meeting space for groups of all sorts, even as a makeshift courthouse); women becoming more aware and frustrated about their inability to contribute to their own well-being (such as not being able to vote, inherit, own property, control reproduction, gain an education or run a business); and competing religious convictions – to name just a few forces. Urban crowding meant that scenes of drunkenness were not confined to private spaces anymore – it became visible. With nowhere but the pub to go, the falling-down drunks littered the streets and were witnessed by children. Sometimes they were the children themselves; one man reported becoming a temperance supporter after having to step over a drunk child to enter a pub. Whether celebrating or escaping, there was a lot of drinking going on.

As social and economic changes were happening so quickly it wasn't initially clear what would grow unchecked and where the state would intervene with rules and regulations. Those in favour of social reform pursued various means to push their agendas. All of these circumstances are relevant to the production of drink maps, why they were made and how they were used. How could any group with a position to assert get their point across without a seat at the governing table? One way was to show – in a concise image – the gist of what would take many pages of written argument. An image such as a colourful map.

DRINKING AT WORK

In the early years of the Industrial Revolution, in the mid- to late 1700s, businesses did not just tolerate drinking on the job and in the fields; they encouraged it and even paid partial wages in beer, cider and whisky. Drinking on the job was normal. 'Small beer' at 2 per cent was nourishing and thirst-quenching, and less apt to bring on gut troubles than the dodgy water, which often mixed with raw sewage on its way to the table before there was a scientific understanding of the potential health hazards. When overzealous sipping took place, it was generally in the privacy of the home. But gradually, as urban centres condensed, drinking became more open. It was (for a while) acceptable at work, in the street and in public houses. Everyone was doing it and at first no one questioned it. This was also the case in the colonies that would become the United States. As the Industrial Revolution was getting started, Benjamin Franklin, then a British colonist but soon to become an American writer and scientist, noticed in the Philadelphia printing press where he worked that men who had beer with breakfast could only manage to carry one heavy printing plate up the stairs at a time, while those who abstained from morning alcohol could carry two. He wrote often about the link between alcohol and reduced productivity, and it was to become an increasingly important issue as the workplace moved indoors and became more concentration and task focused. He considered temperance the most worthy of the thirteen virtues to be pursued in life. Oddly he is also known for his love of French wine – to the point that he powered through his gout and even refused to leave France when his wife became ill and subsequently died. Like many, he struggled with moderation.

As industries began to recognize they could get more out of sober workers they started to impose rules and in some cases side with temperance organizations in the Houses of Parliament. But not all industries, because some found ways to profit from their drunk workers.

Some employees were tethered to their jobs because they had to keep earning enough to pay for their company beer tabs which were run up at the company-owned public houses.

A lot of people regularly consumed beer, including children (who were often also part of the workforce). Families, mostly women, made beer at home. It did not matter that the intricate science of living yeast was not recognized as causing fermentation in beer yet. Brewsters and brewers knew, like bakers, that a little bit of beer from one batch could help the next one get going. Yeast, we now know, is alive. It eats the sugar created in the brewing process and gives beer its fizz and buzz. Back then they were aware of the reliable result – if not exactly why. It also was not understood that the step of boiling water in the beer-making process killed pathogens and made it a healthier beverage to drink than water. But they certainly knew when they didn't feel well and were able to connect stomach aches with drinking water. This is important to under-standing why anyone might have thought it was a good idea to push people towards beer with the idea of getting them away from higher alcohol spirits that were considered more harmful. They already drank it like water, and experienced better health as a consequence.

PRECURSORS TO PERSUASIVE DRINK MAPS

Presenting persuasive information in the form of a map was not new; there were roots in public health. A map published in 1796 showed the spread of yellow fever in New York, and another in 1831 showed the spread of cholera across Asia, Africa and Europe. In England, in 1846 a map made with the aim of solving the cholera puzzle may have in-spired the use of drink maps. Cholera, a disease that mercilessly killed within hours of contraction, was spreading. Dr John Snow, a creative problem-solver, minister and physician, questioned the accepted belief at the time that cholera spread through the air. He suspected that it was instead transmitted through water. To prove his theory he compiled a map of the shared well pumps where stricken victims collected their water in Soho, London. Dr Snow tracked each death, marked them on a map to reveal concentrated patterns, and found they pointed to the Broad Street pump. His map was convincing enough to have the pump handle removed and ended the spread of cholera in that location. One of his points of evidence was that no one in the nearby brewery, which had its own internal water pump, contracted cholera. In spite of Snow being a teetotaller, today you can raise a glass to his map at his namesake watering hole, the John Snow Pub in Soho. Beer in hand, you can also appreciate his other accomplishments such as being an early adopter of anaesthesia for childbirth and administering it to a grateful Queen Victoria for two of her nine children. For the temperance movement, borrowing the idea of visualizing evidence like that on Snow's map made sense. The movement needed to gain credibility beyond the church and its often divisive moral high ground, as many considered teetotallers to be fanatics. One avenue to credibility was to link alcohol consumption to public health and to use a scientific approach to reducing consumption.

As early as 1840, the idea of capturing visual attention with a map to show how closely clumped alcohol purveyors were had been tried out in blues and reds on a map of Newton Heath in Manchester. The use of the words 'drink map' in the title was still more than thirty years away, but a kernel of the idea had been planted on a wall-sized nearly 3 foot by 5 foot map that was initially labelled a 'building map'. That title was crossed out at some point – perhaps its initial use had been outlived – and replaced with *Map of Old Licences at Newton Heath*. Some of the buildings have been hand-coloured and are likely licensed premises, with sixty-five names of public houses recorded. It also has markings of proposed street names in pencil, which were probably added when it was still used as a map of structures. It's not clear if this map was consulted by tax collectors to help them find where to assess licence

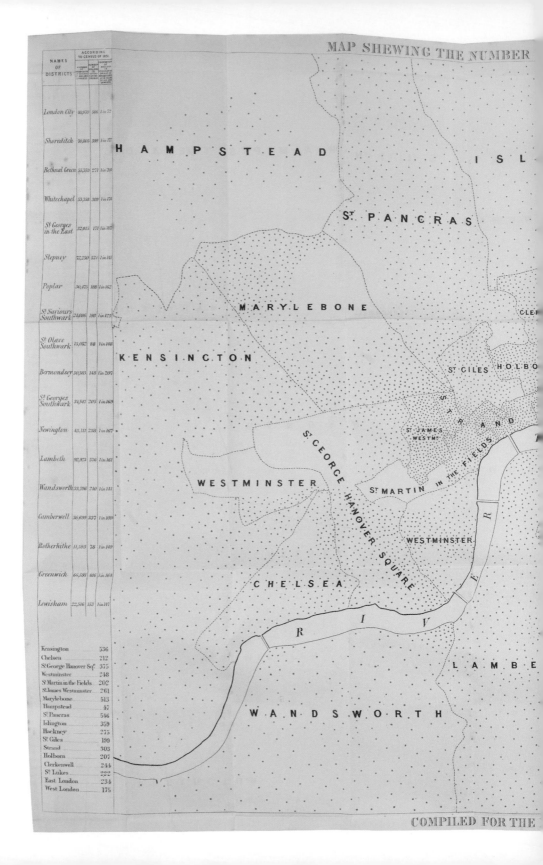

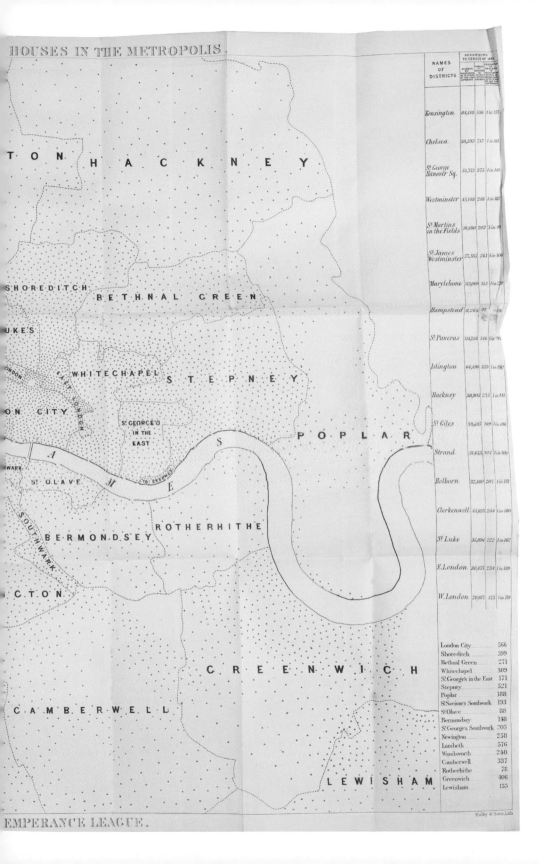

HOUSES IN THE METROPOLIS.

NAMES OF DISTRICTS	ACCORDING TO CENSUS OF 1851		
Kensington	84,410	536	1 in 157
Chelsea	38,393	212	1 in 181
St George Hanover Sq.	55,725	375	1 in 149
Westminster	45,148	248	1 in 182
St Martins in the Fields	18,480	202	1 in 91
St James Westminster	27,553	261	1 in 106
Marylebone	113,009	513	1 in 220
Hampstead	8,264	47	1 in 176
St Pancras	114,254	546	1 in 70
Islington	64,496	359	1 in 180
Hackney	38,905	275	1 in 141
St Giles	38,685	199	1 in 196
Strand	31,673	302	1 in 106
Holborn	32,480	207	1 in 157
Clerkenwell	43,855	244	1 in 180
St Luke	35,894	222	1 in 162
E. London	30,457	234	1 in 130
W. London	20,817	175	1 in 119

London City	566
Shoreditch	399
Bethnal Green	271
Whitechapel	309
St George's in the East	171
Stepney	521
Poplar	188
St Saviour's Southwark	193
St Olave	88
Bermondsey	148
St George's Southwark	205
Newington	258
Lambeth	576
Wandsworth	240
Camberwell	337
Rotherhithe	78
Greenwich	406
Lewisham	153

Malley & Sons Lith.

...EMPERANCE LEAGUE.

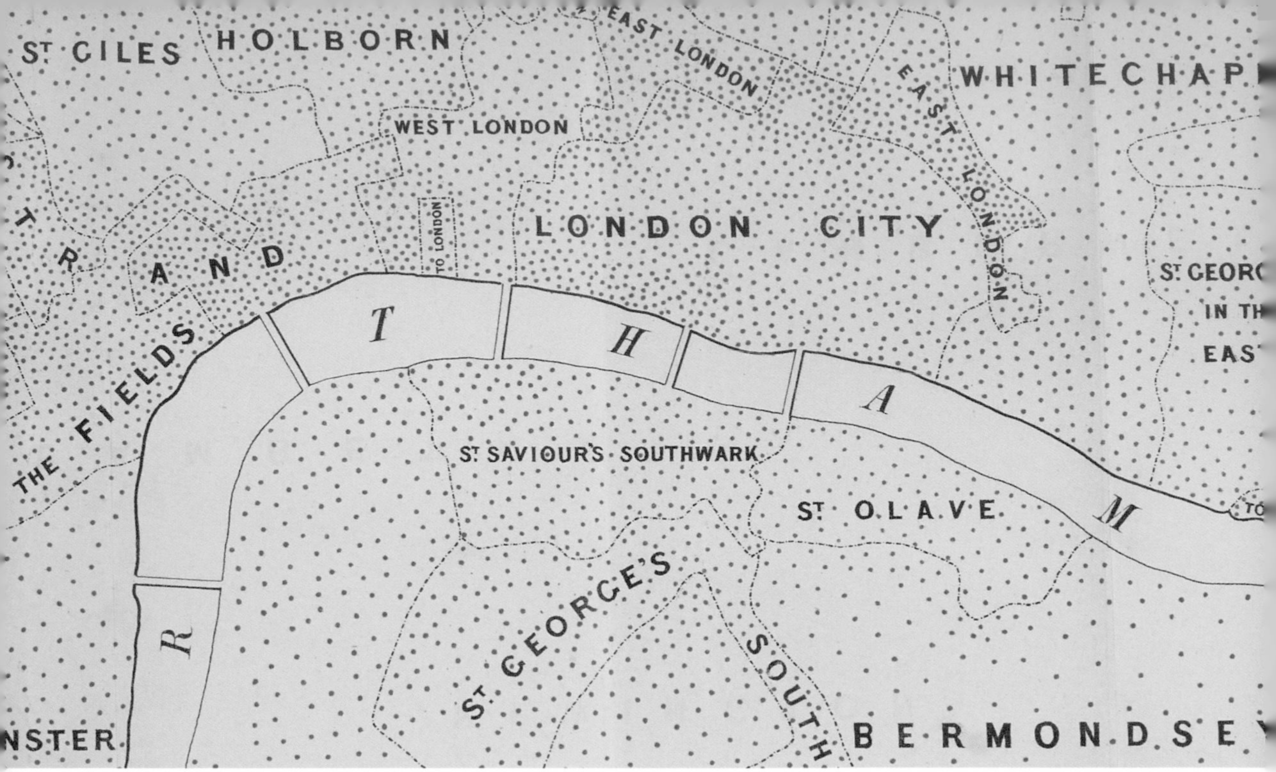
ST GILES
HOLBORN
EAST LONDON
WHITECHAP
WEST LONDON
TO LONDON
LONDON CITY
EAST LONDON
ST GEORG
IN TH
EAS
STRAND
THE FIELDS
T
H
A
M
ST SAVIOUR'S SOUTHWARK
ST OLAVE
ST GEORGE'S
SOUTH
R
NSTER
BERMONDSEY

2 (*and previous spread*) This map was published by the National Temperance League. It included in the margins ratios of public houses per resident of London's neighbourhoods. John Taylor wrote the accompanying plea to reduce the number of places to buy intoxicants. Dots showed patterns of licensed establishments, without streets or geographical context. *Map Shewing the Number of Public Houses in the Metropolis*, 1860.

fees, by police to locate the lively parts of town or by temperance-leaning leaders to note tightly grouped alcohol sales – but licences were certainly central to the map's purpose. This delicate map is held in the Manchester Cathedral Archives.

Another early example of a map depicting drink density was folded into and attached to a pamphlet published by the National Temperance League in 1860. The map, printed by Malby & Sons lithographers, is bound into a booklet and unfolds to an impressive eight times as wide and three times as high as the book itself (FIG. 2). Unlike the Newton Heath map, this map is not a repurposed afterthought. The accompanying publication, written by John Taylor, has its clear motive announced in the title: *Drunkenness, as an Indirect Cause of Crime.* Yet the accompanying map itself is rather opaque. It shows an outline of London's neighbourhoods marked in black overlaid with a spray of tiny red dots. There are no other geographical features at all, not roads nor buildings, only the River Thames. With the exception of a few fields of dot-free areas such as those in Westminster and Poplar there is no obvious connection between the red marks and the locations of the licensed premises they represent. The specks are merely to show quantity. On each side margin is a list of the districts pictured on the map and three columns of information based on the 1851 census: the number of adults living in the area, the number of 'dealers in intoxicating drinks' and a final column combining the information from the first two to show the proportion of drinking places to the population. Statistics being the darling of social reform, this type of ratio would later be used on many drink maps and was a common calculation spouted by those arguing for a reduction of licences. For example, on this map, London City – which takes on a pinkish hue because of so many spots – has a ratio of 1 'dealer of intoxicating drinks' for every 72 inhabitants, while Bermondsey's gently freckled zone has only 1 for every 207. The visual impact of the image dominates the message and leads the viewer to the text to learn more. The four-page paper accompanying the map explains how such an abundance of drinking options came to be, describes all of the problems that can be blamed on this growth and pleads for support of legislation to counter it. It is published by the then-famous temperance publisher William Tweedie, who advertised many titles of anti-drinking books and pamphlets in newspapers and ran a busy shop on the Strand in London. Tweedie also signed this publication as a supporter. John Taylor exhibited a larger version of this same map in the entrance hall of a meeting of the Metropolitan Free Drinking Fountain Association that year. Public drinking fountains were a fairly new phenomenon in the United Kingdom at the time. They were often supported by

A MEETING

OF THE MEMBERS AND FRIENDS OF THE

National Temperance League

WILL BE HELD IN THE

LARGE HALL

OF THE

YOUNG MEN'S CHRISTIAN ASSOCIATION, ALDERSGATE STREET,

ON

TUESDAY EVENING,

JULY 26th,

TO TAKE INTO CONSIDERATION THE SUBJECT OF

DRINKING FOUNTAINS

IN THE METROPOLIS.

WILLIAM JANSON, Esq., GEORGE CRUIKSHANK, Esq., T. B. SMITHIES, Esq., Dr. ELLIS, of Sudbrook Park, Mr. J. P. PARKER, EDWARD THOMAS WAKEFIELD, Esq., Hon. Secretary to the Fountains Association, and other Gentlemen, will address the Meeting.

CHAIR TO BE TAKEN AT EIGHT O'CLOCK.

3 Drinking fountains were competing with beer as a means of refreshment, and were intentionally placed near public houses. Advertisement for a meeting to discuss drinking fountains in London, organized by the National Temperance League (undated).

temperance groups and strategically placed near public houses in an attempt to provide an alternative refreshing beverage. At Taylor's drinking fountain meeting it was noted that a similar map had been shown to magistrates at a water fountain support meeting in Glasgow with the effect of preventing an increase in licensed public houses.

That same year, in 1859, a parish map of Eastbourne had been traced and affixed with the locations of public houses. It was then submitted to local authorities to show how little distance there was between drink shops in an attempt to persuade them to stop granting more licences. The map does not survive, but it made an impact.

A fourth map used in the early days of image propaganda is one of Aberdeen published in 1872 (FIG. 4). This *Map of the City of Aberdeen shewing the numbers & position of the Licensed Grocers & Spirit Dealers* prefaced the annual report of the Aberdeen Temperance Society. To understand the possible message one has to read between the legend lines. Only two types of licences are shown, grocers are represented with an outlined empty box and all others, including licensed hotels, inns, spirit dealers '&c', are shown by shaded boxes. Both types are concentrated in the centre of town near the markets and railway station, and then disperse in ripples though often skipping entire areas. The point, considering how the markings were divided up, appears to be that when comparing drinking density there is a disproportionately high availability of take-away intoxicants.

All four of these maps, while different in size, location, content and publisher, are considered drink maps because of their intention. Neither a speech nor a written list could have had the same immediate value as a glance at a map that was both personal and communal. Everyone could recognize their own personal place on it while at the same time spatially comprehending how each position related to others. The maps, regardless of whether their viewers considered them to be good or evil, gave them their drink bearings.

THE TEMPERANCE MOVEMENT
GAINS MOMENTUM

Temperance organizations popped up all
over the world in the early 1800s. Early
anti-drinking groups were established in
Sweden in 1819, Italy in 1830, Denmark in
1840 and Poland in 1844, to name a few.
Many of them asked their members to sign
a pledge promising not to drink spirits.
In Preston, cheese seller Joseph Livesey
noticed that those who refrained from hard
liquor but continued to drink other alcohol
tended to backslide into intoxication and he
famously signed a pledge of *total* abstinence
in 1832. The concept of shunning all boozy
beverages, known as teetotalism, was not
new. But it gained in popularity after this
widely publicized event. The battle between
abstinence and moderation would rage on
and eventually became an insurmountable
obstacle to the success of the British temperance movement's goals. In many areas, such
as health, social acceptability and religion, it
continues today.

4 This map was published by the Aberdeen Temperance
Society showing the increased concentration of dealers
of intoxicants approaching the city centre. *Map of the City
of Aberdeen shewing the numbers & position of the Licensed
Grocers & Spirit Dealers*, 1872.

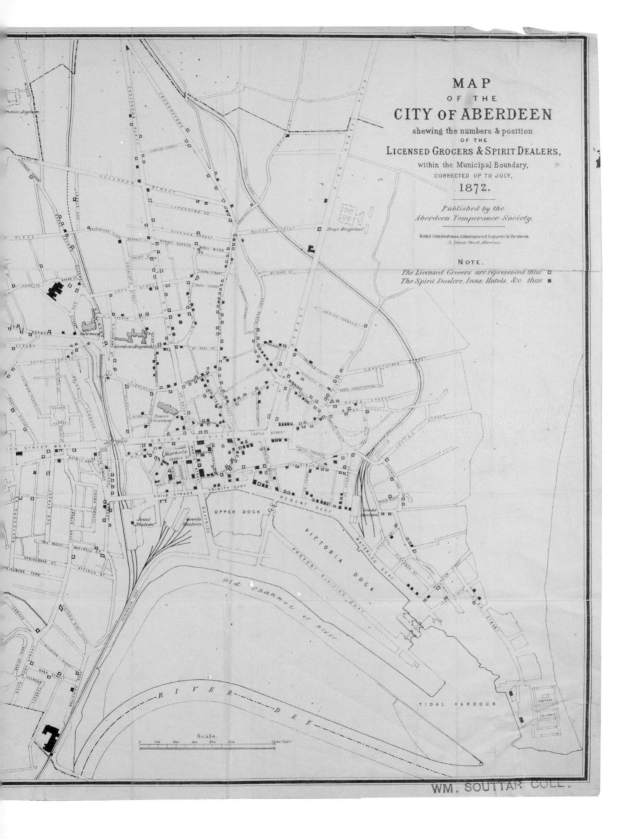

MAP
OF THE
CITY OF ABERDEEN
shewing the numbers & position
OF THE
LICENSED GROCERS & SPIRIT DEALERS,
within the Municipal Boundary,
CORRECTED UP TO JULY,
1872.

Published by the
Aberdeen Temperance Society.

Keith & Gibb, Draftsmen, Lithographers & Engravers to The Queen.
3. Queen Street, Aberdeen.

NOTE.

The Licensed Grocers are represented thus □
The Spirit Dealers, Inns, Hotels, &c. thus ■

Around the world, new laws attempting to curb alcohol consumption were implemented in Canada, Australia and Switzerland. A couple of examples of the breadth of the drink problem phenomenon can be seen in what is now Germany and in France. In Germany, the drink problem was labelled the 'Drink Question' and studied with a scientific approach. There was serious debate about how much beer was too much at work. From 1837 to 1846 the number of temperance societies in Germany grew exponentially from a handful to over 1,000. The drinking water supply was questionable in Germany just as it was in the UK, so beer was a daily beverage for all ages. German workers ate five meals a day – three during work hours and usually in taverns that offered beer and schnapps. In the evenings, taverns also hosted meetings for clubs – so drinking was an expected 'thank you'. Essentially imbibing was non-stop. Towards the end of the century restrictions were implemented limiting the drinking age to sixteen, not allowing sales to anyone convicted of drunkenness in the previous three years, fining beer houses for serving obviously intoxicated people and compulsory placement of multiple offenders in inebriate homes.

France, like England, at first looked to puritan American temperance reformers for ways to approach their growing number of inebriates, paupers, poor and undernourished. In 1875 French absinthe drinkers were guzzling it in record numbers. It became even more popular the following decade when the grape crops failed and absinthe became more readily available than wine. But while there are records of the French considering an alcohol ban similar to the strict American laws, it never got off the ground. It remains a country of moderate yet daily drinkers, and generally of wine over spirits and beer.

Other countries were dealing with a surge in alcohol intake as well. Today it is hard to conceive of such a passionate stance against alcohol. If anything, there is now more pressure to drink than there is acceptance of those who choose not to, although recently this has started to shift again. What was all the fuss about – what's wrong with people enjoying a drink or two? Part of the tale is the sheer amount of alcohol being consumed around the world. There were years in the UK when an entire third of the annual tax revenue came from alcohol, compared to today being closer to 4 per cent. New public-house openings were nearly out of control with no end in sight. Families were completely reliant on the adult male – the husbands and fathers – for money. So if a man drank his pay on the way home, the family literally starved. Companies often paid their employees on Fridays from the very pubs they owned, where the workers then spent their celebratory new cash. Women turned to the churches for charity and safety. But churches also had limited resources, and therefore had an interest – moral and financial – in slowing the mess left in alcohol's wake.

It's possible that John Snow's cholera map inspired another physician to use a map to help him argue against the ills of drinking alcohol. Dr Thomas Nichols and his wife Mary were already an infamous couple, having made the British papers in 1857 when they renounced the free-love commune they themselves had founded in America to be baptized as Christians. As devoted abolitionists, they fled the USA shortly thereafter to escape the unrest of the coming civil war and relocated to London. For the next several years they published a newsletter titled *The Herald of Health* and lectured about the benefits of vegetarianism, medicinal water treatments and teetotalism. Dr Nichols publicized one of his lectures in the temperance publication *Alliance News* on 15 December 1877, and in his article described a giant map he had made as a backdrop for his lectures. Picture this: a map extending from floor to ceiling in an 8 foot by 8 foot square showing just a few streets of a city. It was sometimes called 'Nichols's Sensation Map'. He wanted those at the back of the hall he was speaking in to see it clearly. The map showed just a half-mile area of London, so the scale was quite grand, on which he had 'coloured in jet black every public-house, brewery, and

5 A selection of Pledge books, *c.*1890. Pledges were signed in front of witnesses and were reminders and portable proof of the carrier's commitment to sobriety. The *Personal Temperance Pledge Book* (*c.*1890) was used both to record who had signed pledges and to give the signatory a card to carry as evidence and reminder of their commitment to sobriety.

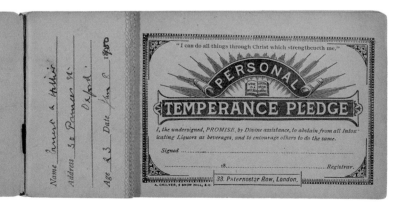

distillery'. He referred to the 276 black dots, which ranged from 1 to 4 inches, as an 'infernal constellation', thereby missing an opportunity to conjure a scarier comparison like cholera, the plague or at least a rash. He was a big believer in a regulatory approach to reducing alcohol (as promoted by the United Kingdom Alliance (UKA), a powerful temperance organization founded in Manchester in 1853 and publisher of the *Alliance News*). He compared the work of some temperance groups still trying to get individuals to sign pledges in front of their peers as being as pointless as 'trying to dip out the ocean with a spoon'.[1]

Nichols offered to take his lecture titled *One Half-mile Square in the Heart of London* anywhere *Alliance News* readers invited him, and also encouraged others in the UKA to make similar maps of the most densely pub-pocked portions of their districts. He did not call his map a drink map, but it surely was one in tone and mission. His instructions referred specifically to *licensed* houses – using their legal status as opposed to simply saying beer shops or public houses. This was consistent with the overall original intent of the UKA's founding to attack drinking through law and regulation instead of moral authority. Dr Nichols's lecture was later published and provides a narrative where he links the space on his map to human scale. In his instructions he insists that those looking at the map visualize themselves in the three-dimensional vicinity it represents. He encouraged viewers to visit the actual street, to '[w]alk slowly, and look up the alleys on either side'[2] of Drury Lane.

THE VICTORIAN WORLD OF MIND-ALTERING OPTIONS

A typical image promoted by proponents of temperance was that of people of all ages in rags stumbling around drunk or passed out in the street – which police reports at the time confirm was often actually the case. But it wasn't just the poor and the new urban working class that were tempted by a tipple. Everyone was drinking. Those with

ONE HALF-MILE SQUARE

IN THE

HEART OF LONDON.

This Map is reduced from the Great Map of Sixty-four Square Feet, scale of 16 feet to a mile, drawn by Dr. Nichols, to illustrate his Lecture on "One Half-Mile Square in the Heart of London." It is divided into sixteen squares of forty rods, or one furlong, each, and exhibits every Public House, and the Distillery, Breweries, Theatres, and Music Halls, but not the Beer Houses or licensed Grocers and Wine Dealers.

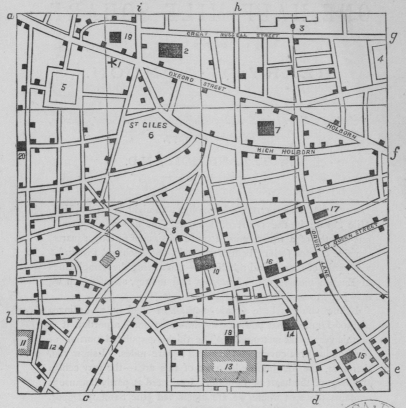

REFERENCES.

a. Oxford Street, leading to Hyde Park and Tyburnia.
b. Cranbourn Street, Long Acre, and Queen Street.
c. St. Martin's Lane, leading through Seven Dials and St. Giles.
d. Wellington Street, Bow Street, Endell Street, &c.
e. Drury Lane and Museum Street to the British Museum.
f. Holborn, leading across the Viaduct to the City.
g. Great Russell Street, leading to Tottenham Court Road.
h. Line of Gower Street, from Waterloo Station to Euston.
i. Tottenham Court Road, crossing Oxford Street, to Crown Street, and projected new avenue to Charing Cross.

1. Sanitary Depôt, 429 Oxford Street, Salsbury Hall, Reading Room and Library. 2. Great Horse Shoe Brewery of Meux & Co. 3. British Museum. 4. Bloomsbury Square. 5. Soho Square, 6. St. Giles's Church. 7. Bloomsbury Distillery. 8. Seven Dials. 9. Newport Market. 10. Great Brewery in Castle Street. 11. Leicester Square. 12. Alhambra Theatre. 13. Covent Garden Market. 14. Covent Garden Theatre. 15. Drury Lane Theatre. 16. Queen's Theatre. 17. Middlesex Music Hall. 18. Evans' Music Hall and Hotel. 19. Oxford Music Hall. 20. Little Royalty Theatre.

LONDON: NICHOLS & CO., 429 OXFORD STREET, W.

PRICE ONE PENNY

comfortable means could drink privately at home or at their clubs, where the worst result of imbibing too much was an embarrassing story the next day instead of losing money meant for the baby's shoes. For those living in squalid and cramped abodes, going to the public house was as much for the improved surroundings as for the alcohol. On drink maps that distinguish different types of licences, it's easy to pick out the wealthy neighbourhoods even if you're not familiar with the town or city because not only are there no bright dots showing alcohol outlets in rich residential areas, but the dots on the fringes of these places usually indicate more off-licence shops than public houses; outlets for servants to conveniently access bottles of wine and spirits for the cupboards and cellars of well-stocked homes.

To say that people were drinking a lot doesn't paint the whole picture. Cocaine was not stigmatized yet, nor illegal. It was consumed without shame, available without a prescription, and took off in the 1850s. In spite of the deceptively simple story shown on drink maps of places overrun by one type of demon, the intoxicant problem was considerably more complex. Ultimately, three things inspired the initial production and use of drink maps in an attempt to influence the rules related to imbibing for everyone: unprecedented public drinking in crowded areas (as we have seen), legislation that backfired causing more drinking than it deterred (as will be described next), and the addition of new legal strategies to augment organized social-pressure tactics already in use.

6 This map is a reduced version of an 8 × 8 foot map that Thomas Nichols used as a backdrop for his lecture of the same title. It shows a whopping 276 places to find booze in Central London. *One Half-mile Square in the Heart of London,* 1878, first page. Oxford, Bodleian Library, G. A. Lond. 8o 441 (3).

WHITBREAD'S

PER DOZEN IMPERIAL PINTS		HALF PINTS
LONDON COOPER		
HALF CROWN STOUT		
FAMILY ALE	**2/6**	**1/6**
INDIA PALE ALE		
LONDON STOUT	**2/9**	**1/9**
EXTRA STOUT	**3/-**	**2/-**

ASK FOR
WHITBREAD'S
LONDON STOUTS, COOPER
& ALES IN BOTTLE

Sold by J. STRONG,
Farmers' Arms,
WALLASEY.

EYRE & SPOTTISWOODE. LITH LONDON

CHAPTER TWO

THE DRINK TRADE
& THE LAW

While temperance organizations and lawmakers were trying to find a
way to balance the interests of the public with those of the drink trade,
beer was improving. Many technological advances came about in the
1800s that aided the making, packaging, stability and distribution of
beer. And, most importantly, the taste was getting better.

Few have lived the adage that necessity is the mother of invention
like the brewers of nineteenth-century England. High and fluctuating
taxes, various wars, flip-flopping laws and unpredictable ingredient
availability – all forced breweries to be innovative at short notice.
The three major London brewers in the early 1800s were Whitbread,
Barclay Perkins and Truman's; close competitors were Charrington and
Courage. One particular demand that has left an enduring mark on
the history of beer was the need to brew for the British troops in India.
A pale ale was sold by George Hodgson to the East India Company
not because it was the best, or the first, but because his docks were
more convenient than the larger brewers to East India Company's
docks making delivery easy. And so began the India Pale Ale, today

7 Advertisement for Whitbread's Ale and Stout, including India Pale Ale and Family Ale, c. 1860.

known as IPA. It was a hit, and Hodgson had an exclusive contract for a while. Historians believe that both highly hopped porter and a golden 'October beer' were sent to India, with the upper-class officers being more accustomed to the lighter and more expensive brew with the darker beer being designated for the workers. It was the paler beer that caught on from that era and enjoys popularity today.

Soon Hodgson, or more accurately his sons, became greedy and tried to set up their own shipping company to circumvent paying for the voyage to India once the quality and demand for their beer had been established. A variety of sketchy tactics were employed until the East India Company decided to approach another brewery, Allsopp's in Burton upon Trent, and asked them to mimic the beer so they could undercut Hodgson's. And so an Allsopp brewer tried to copy Hodgson's beer, meaning he tried to make a beer that would be lighter in colour and not as sweet as their usual brew. They did not have the chemical understanding yet to realize that the gypsum in the local water, some-times referred to as hard water, was better for the conversion of grains to sugar, for superior colour and for a more elegant hop expression. One thing was certain – the Burton water made a better-tasting beer. It was a huge success, and not just in India.

Barclay Perkins, based in London, was the largest brewery in Britain in the 1850s. It was well known enough to be something of a tourist destination and gave perhaps the first brewery tours. The largest breweries were also powerful in government. Charles Barclay of Barclay Perkins was a Member of Parliament, as were Sir Thomas Fowell Buxton of Truman's, Samuel Charles Whitbread of his family's namesake brewery and William Wigram of Meux Reid. Of the brewer MPs, most didn't just represent brewing interests, but big business interests

8 Advertisement for Allsopp's lager highlighting its drinkability around the world and in every climate, c. 1900.

– two-thirds of them led companies with values of over £500,000. Brewer MPs numbered fifty at their peak in 1886 with a significant number coming from London. Temperance sympathizers certainly had their share of MPs as well, but not nearly to the same extent. Charles Buxton, a brewer MP who fought the spirit trade and was mentioned on the Drink Map of Southampton (see p. 45), was not the only politically powerful brewer with a mixed agenda that sometimes aligned with temperance groups. The drink trade's lobby, such as it was, was known for infighting – small breweries versus large ones, urban versus rural, brewers with tied houses (pubs that exclusively sold one brewery's beer in exchange for financial backing) versus independent public houses. Eventually a single entity called the National Trade Defence Fund tried to be an umbrella organization to any business related to alcohol, including hop growers and glassmakers, to mixed results. The lobby itself was not that important to national-level efforts given the trade's representation in Parliament, and the drink trade was missing the grassroots side of public persuasion. This was in contrast to the temperance force, which disseminated their message on a micro-local level, one mapped town at a time. Historians have tried to explain why the drink trade only made half-hearted efforts to organize more customer-level counter-publicity. There were certainly trade associations, and the brewers had a guild, but it never turned into the well-oiled lobbying machine one might have expected from the funded business side of the drink equation.

LATHER STORIES

Yeast was identified as causing fermentation in the 1830s. It was certainly already understood that yeast was important to making beer, but that fermentation was an active *process* involving a living entity, not just a chemical reaction, was the new part. Still, knowledge was slow to spread. Louis Pasteur visited breweries in 1871 and showed surprised brewers that he could see which of their batches were infected or had gone off by looking at them under a microscope instead of tasting them. The 'why' was unveiled. It wasn't that breweries didn't already have chemists. Bass; Truman, Hanbury & Buxton; and Worthington all had laboratories since the introduction of the sacchorometer, a device used to measure how much fermentable sugar was in a given liquid. Pasteur's namesake process called pasteurization was applied to beer shortly thereafter. Through heat, it destroys many of the microorganisms that can spoil beer, allowing for a superior flavour and longer storage. All of these improvements dramatically raised brewers' abilities to control the quality of their products and made beer desirable to more people.

Temperance supporters were also paying attention to the ingredients of beer. Joseph Livesey, the Preston man dubbed the father of teetotalism because of his well-known total abstinence pledge and ceaseless evangelizing on the topic, asked in an article 'how many wife beatings … may proceed from a single field of barley?'[1] This was someone who clearly understood that brewers' logs always noted the £ per malt yield measured for every batch of beer, and he was pointing to what he saw as another cost. As a youth, he had attended a sermon about equality. The dinner afterwards required a fee, and he could not afford it. He scolded the drinking attendees about words versus deeds, and in that moment discovered both his knack for sharp articulation under pressure as well as his awareness of unfairness. He helped start a reading room that loaned newspapers to the public, and then a library. He edited an anti-drinking journal titled the *Temperance Advocate*, which sprang from his social reform publication in 1832. He edited a journal entitled *The Staunch Teetotaler* 1867–68; the numbers were collected into a bound volume in 1869 and published by the renowned temperance publisher William Tweedie. Livesey's funeral in 1896 was a parade of carriages ferrying the who's who of temperance.

Like Livesey, many temperance advocates began their rallying speeches with their own personal stories of a flashpoint incident that turned them against drink. And when that storyteller was from a family that owned one of the largest breweries in England, and his conversion cost him his family fortune – the story went Victorian-speed viral. Frederick Nicholas Charrington was just twenty years old in 1870 when he was walking home from his job at his family's massive brewery in Mile End, London. He saw a woman with small whimpering children pulling at her skirts; she put her head inside the door of a public house to find her husband. When the man emerged she asked him for some bread money because the children had not eaten that day. The man knocked her into the gutter and went back inside. A shocked young Frederick looked up to see – in large fancy letters above the door – Charrington. The pub was a tied house of his family's brewery. After that, to his family's dismay, Frederick gave up his shares and became an ardent temperance reformer for the rest of his life. He established that same year a home for upper-class inebriates on Osea Island in Essex connected to Tower Hamlets Mission. He renounced his shares in the brewery, estimated to be worth over £1,000,000, which says a lot about the money being made by large brewers of the day. In telling the story he would say that what knocked that woman down also knocked him out of the drink trade.

INSIDE PUBLIC HOUSES

The gradual shift of where people were drinking may not have been noticed by people going about their daily lives, but the new spatial concentration was immediately visible on maps. At the start of the century in rural areas, people drank at inns when they had to rest their horses on long journeys. It was where they found their beer, their bread and their bed. The switch from horse-powered coaches as primary modes of long-distance travel to trains also meant that cities became more crowded as families followed the work, and more options for alcohol followed the money. From fancy gin palaces which oddly flourished long after the gin craze to the sawdust and spare atmosphere of the beer house, urban areas offered a bounty of booze. Many were plonked in the midst of squalor. In nicer areas, public-house interiors were more refined, and tried to appeal to all classes of customer, with different rooms for each of them. There was usually a taproom where everyone passed through and some stayed to drink standing up. In early days a barricade in the form of a rail barred customers from helping themselves to the taps, and eventually became known as a bar. Today the legacy of this arrangement of different rooms in public houses is widespread in the UK, but the rooms have different names in different regions. In general they were more like many small living rooms and allowed the classes to self-separate. The 'tap' became the name of the general room; posher customers could move to the parlour. Regardless of location, it was a male-dominated community.

THE LEGAL LANDSCAPE

When it comes to alcohol laws today, everyone is more of an expert than they realize. The average person knows how late the pub is allowed to stay open, the legal limit to drive a car, the minimum drinking age, if and where beer can be purchased on Sundays, whether it's okay to drink openly on public transportation, at work or in a park. These laws are known; whether adhered to or circumvented is a calculated choice because most people are also aware of the punishment for violating them. But in the 1800s none of the rules had been sorted out yet and fighting over what would become the regulations was a major focus of the era. There was not yet an infrastructure in place – either physically or legally – to handle the repercussions of a rapidly changing world.

(overleaf)
9 *Beer Street* by William Hogarth, etching and engraving, 1751.
10 *Gin Lane* by William Hogarth, etching and engraving, 1751.

BEER STREET.

Beer, happy Produce of our Isle
Can sinewy Strength impart,
And wearied with Fatigue and Toil
Can chear each manly Heart.

Labour and Art upheld by Thee
Successfully advance,
We quaff Thy balmy Juice with Glee
And Water leave to France.

Genius of Health, thy grateful Taste
Rivals the Cup of Jove,
And warms each English generous Breast
With Liberty and Love.

Design'd by W. Hogarth.   Publish'd according to Act of Parliament Feb. 1. 1751.   Price 1.s

GIN LANE.

S. GRIPE PAWN BROKER

GIN ROYAL

KILMAN DISTILLER

Gin cursed Fiend, with Fury fraught,
Makes human Race a Prey;
It enters by a deadly Draught,
And steals our Life away.

Virtue and Truth driv'n to Despair,
It's Rage compells to fly,
But cherishes with hellish Care,
Theft, Murder, Perjury.

Damn'd Cup! that on the Vitals preys,
That liquid Fire contains,
Which Madness to the Heart conveys,
And rolls it thro' the Veins.

Publish'd according to Act of Parliam. Feb. 1. 1751.   Price 1.s

Industries needed a lot of workers and fast; they hired with little regard for anything other than efficiency and cost, which launched an era of labour exploitation. Meanwhile, sluggishness at work and absenteeism connected to overindulgence by workers hit the pockets of industry owners and grabbed the attention, for the first time, of property-owning (voting) men about the drink problem.

The late 1700s were known as the days of the 'gin craze' because so many people were consuming high-alcohol spirits. The trend had been quelled for a while through high taxes, but by the 1820s, thanks in part to a significantly reduced spirit duty, the drinking of copious amounts of gin continued to increase and public drunkenness and poverty escalated as wages tended to get swallowed on the way home. The legislature was desperate to take swift and broad action to stifle the resurgence of spirit quaffing that was on par with the gin craze. It was widely believed that if people switched to drinking beer instead of gin, everyone would be better off. At the end of the previous century William Hogarth had provided visual support of this notion in two images meant to be compared: *Beer Street* and *Gin Lane* of 1751 (FIGS 9 & 10). The beer drinkers enjoy an abundance of food and mirth – they are working, flirting and happy; while the gin drinkers are trouble – fighting, clumsy and miserable. The introduction of a new law – the Beerhouse Act – was meant to provide the solution.

THE ALEHOUSE ACT

Before the passage of the Beerhouse Act was the enactment of a procedural mechanism: the Alehouse Act of 1828. This was the first in a series of licensing laws spanning from 1828 to 1886 that went through modifications, additions and amendments. Think of it as the first thread that quickly became a tangle of knots. This initial Act provided a structure to regulate how beer was sold, including establishing an annual licensing day in each town called a 'brewster session'. Just once a year the licences were available, and if denied (which could happen simply by not showing up) applicants had to wait another whole year to apply, and were forbidden from selling any alcohol even if they had done so the year before. On the designated day each year, the local inn, grocery and alehouse owners presented their applications to a group of magistrates for the right to sell alcohol for the next year. At these brewster sessions or 'licensing days' applications were submitted along with a fee. At first the licence seekers had to apply in person, but the granting of licences – especially if a business had been granted one the previous year – became fairly automatic, and the physical presence of the applicant was often not even required. Objections to the granting of licences could be lodged within a set amount of time in advance by anyone with an interest and a shilling. Most objections were made by police officers (who had to deal with high incidences of drunkenness and drink-related crimes) and by the magistrates themselves (usually when a drinking establishment was proposed in their own neighbourhoods, or motivated by the magistrates' views on temperance). Objections had to be based on evidence relating to a moral deficiency in the applicant or other reason related to indecency or dishonesty, and on the 'needs of the neighbourhood' (although this was not defined). While temperance advocates were allowed to serve as magistrates on licensing days, magistrates with ties to the drink trade were not permitted to rule on licensing applications. The rationale was that brewers, public house and grocery owners, hop growers and maltsters had a financial stake in the decisions.

THE BEERHOUSE ACT: INTENTIONS BACKFIRE

The stated logic for the new Beerhouse Act was that a wider availability of beer would be a deterrence to drinking gin. (The word 'gin' at the time was frequently used to mean all distilled spirits.) It came from a sincere belief that the new legislation would change the course of drinking towards sobriety. The proposed Act would allow anyone to sell beer almost as easily as hanging out a sign. It was discussed and debated, and ultimately became the winning argument for the Act's passage into law. If people only had more places to get beer, they would surely steer clear of gin.

The result was legislation which allowed anyone to sell beer who paid a nominal fee. Truly anyone. Without magisterial scrutiny or regulation, and operating outside of the licensing framework, it prompted an explosion in the number of beer houses. Even before the Beerhouse Act, most people already brewed beer at home. It was easy to make, although no doubt of varying quality and tastiness. After the Beerhouse Act, beer finally cost less and was more widely available than gin, and it became ubiquitous. It did not take long for the law to be blamed for promoting, not deterring, drinking – undermining its original intent. And yet repeal was elusive because of timing, power shifts and more urgent priorities.

Forty thousand new beer houses opened in the first two years of the Act's passage alone, and continued to open apace. Liverpool saw fifty new houses open every day for weeks. As one citizen put it, 'The new Beer Act has begun its operations. Everyone is drunk.'[2] Complaints by the public about the sheer visibility of so many outlets made the newspapers – including an account of a father arguing that there were so many pubs on his street there was no other way for his small children to walk home without seeing the visible effects of drink; he worried they would grow up thinking it was normal and acceptable.

Watching the booming beer trade in horror, temperance and teetotal groups ramped up their activities. Membership and donations increased. New organizations were formed, including one in Manchester called the United Kingdom Alliance in 1853, which became the largest and most influential temperance organization in Britain with chapters all over England and Wales. The UKA embraced the idea of disseminating drink maps to shock people by showing them at a glance just how many places there were to buy alcohol. Their tactic was to spell out on the maps themselves the case against drinking and the meaningful action citizens could take directly, thereby spreading the word both visually and literally with the ultimate aim of persuading magistrates and lawmakers.

FROM PEER PRESSURE TO AUGMENTED LEGALITY

All legislation is valueless unless it removes the temptation to drink. There is a law against selling drink to anybody under 16. I would just increase that figure and say 85 years of age, and the law would be all right.

SIR WILFRID LAWSON, 1878[3]

There is a street in Manchester named after the US state of Maine, 'Maine Road'. The honour is based on the United Kingdom Alliance's appreciation of the approach of the mayor of Portland, Maine, to ending rampant drinking: controlling or banning it entirely by law. Portland's sober leader, Neal Dow, had been the president of his local temperance organization for less than a year when he became mayor of the largely working-class port city. He grew up a strict Quaker and despised not just alcohol itself, but those who drank it. He openly and vocally blamed Irish and German immigrants for the poverty, criminal behaviour and other vices he associated with drinking. His method of legislative regulation, as opposed to church-based moral pressure, became known as the 'Maine Law'. It was a statute that forbade the sale of alcoholic beverages with a few narrow exceptions, including medicinal use. It passed in 1851. A few years later, some angry and thirsty citizens believed (correctly) that Dow had stashed $1,600 worth of alcohol in the basement of Portland City Hall (about £38,000 today). Although it was meant for the permitted exception of medicinal dispensary, they angrily descended upon the government building. Dow told them to leave, and when they refused he ordered the unarmed group to be fired upon. One died and others were wounded. The disproportionate response and relentless public outcry led to the Maine Law being repealed by 1856, and also led to Dow's international disgrace. But, in the end, he still won. Prohibition was written into the state's constitution in 1885. Maine was the first state in the United States to embrace the full prohibition of the production, transport and sale of alcohol. It would be another thirty-four years before the entire United States endured nationwide Prohibition by federal constitutional amendment.

Before all the drama, the Maine Law's initial success got the attention of other countries grappling with the 'drink question', including France, Germany and Scotland. Scotland in particular, already having a robust temperance-minded community and organized societies like the Glasgow and West of Scotland Temperance Society founded by John Dunlop in 1829, was hitting a wall with its one-prong approach to the problem, which was essentially religious guilt or piety (depending on the level of desire for a drink). Scotland embraced Dow's radical tactics of reducing drinking through state-driven restrictions and liquor trade regulations instead of appealing to individuals' morality and willpower. Dow was so admired by like-minded people in the UK that he was invited to England to go on a speaking tour of sorts and to meet with temperance group leaders; he ultimately spent over ten years on multiple trips to spread his word.

With a nod to Dow, solving the drink problem by legally removing its temptation became one of the founding principles of the United Kingdom Alliance when it was formed a couple of years later in 1853 with the stated mission 'to *outlaw* all trading in intoxicating drinks' (emphasis added). At first, UKA membership was open to drinkers. Some early leaders and members of the UKA were also Members of Parliament, and they used their seats to influence legislation. The internal structure of the UKA was geared towards doing just that. Each UKA town, city and regional branch had at least one member designated as a government affairs liaison, meant to interact with and influence government – essentially, a lobbyist.

The most prominent and vocal member, and eventual UKA president, was Liberal Party MP of Carlisle Sir Wilfrid Lawson. He was born just a year before the Beer Act of 1830, and therefore entered a world of widespread alcohol consumption. From his early religious teachings, and coming from a family of temperance advocates, he was groomed against drink from the start. And while he was anti-opium, anti-prostitution and anti-gambling, he was also known for being the life of the party – at least in terms of humour. He wrote a memoir of his participation in the major events of his time as a Member of Parliament with his personal views on everything from British involvement in Egypt, India and Ireland to barrel-organ street performances. He also assembled collections of his thoughts, jokes and sarcasm such as *The Wit and Wisdom of Sir Wilfrid Lawson* (1884), *Wisdom Grave & Gay* (1889) and *Cartoons in Rhyme and Line* (1905), which were popular at the time. He seemed to take pride in getting lambasted in *Punch* cartoons.

The UKA buttressed its efforts by publishing a weekly newspaper titled the *Alliance News*. It was used to strengthen strategic connections and promote the effective dissemination of information, and was a friend to all temperance organizations. It used its platform to share model language to make it easier to oppose licence applications, templates for anti-drinking letters to MPs, sample petitions, information about political candidates' positions on various pieces of anti-drinking legislation, and a comprehensive calendar of meetings and activities for all temperance organizations – not just its own. The UKA was generous with information and reprinting of articles from other newspapers that featured temperance success stories. It became the mouthpiece of the movement.

The drink trade had its own methods of print communication, including the *Brewers' Guardian* with a masthead announcing that the paper was 'devoted to the protection of brewers' interests in licensing, legal and parliamentary matters'; and tellingly underneath 'The organ of the country brewers'. This distinction between country and city brewers would become a deepening chasm throughout the century, although the *Brewers' Guardian* developed into an international magazine. In 1871 it was the first to report on the visit by Louis Pasteur to Whitbread brewery in London. The *Brewers' Guardian* only recently suspended production, with its last print issue in 2011. Its online format came to a close shortly thereafter. In its heyday, there were at least forty different drink-trade newspapers, some regional such as the *Northern Brewers' & Victuallers' Journal* and the *Scottish Wines, Spirits & Beers Trades Review*; some were specific to a type of licence, such as the *Licensing News & Public House Trades* and the *Licensed Wine Trade Circular*; some even for the type of worker, such as London's *Barman & Barmaid* of 1879.

Lawson was impressed by the American concept of reducing alcohol intake by legally banning its availability. Yet he did not seek complete prohibition all in one law, and instead pursued a course of chipping away from multiple directions. Over the years as an MP, he introduced legislation on several alcohol-related initiatives. Few passed, but his reputation grew and he gave voice in the legislature to a growing anger towards drink and the related expensive problems representatives were being asked to solve by their constituents. Although he did not introduce legislation banning the sale of alcohol, he instead drafted laws that would restrict the availability of it and also to expand the power of people to decide what to permit in their own surroundings. Over the years his position on drinking, even though it did not change, went from being seen as radical to being reasonable. Generally speaking, at the time the Liberal Party was as aligned with temperance as it was with most positions meant to protect the public, while the Conservative Party tended to protect business interests and was supported by the drink trade, although there were subtleties within the parties that were not so black and white. Lawson supported bills restricting hours, denying payment to establishments that claimed they should be compensated when they were refused licences, and attempted to give local people more of a say about when licences were to be granted in their own backyards. For example, in 1863, in a speech in support of a Sunday closing bill introduced by the MP for Hull, Lawson argued:

> I think the keeping of Public Houses open on Sunday is about
> the most ghastly-comical thing in all our system of anomalies and
> absurdities. The day is divided between the service of the Almighty
> and the worship of Bacchus – the latter being far the more popular
> and patronized.[4]

Many schemes were proposed by different advocates. In 1876 Joseph Cowen, MP for Newcastle upon Tyne, introduced a bill to bring the drink trade under municipal control, meaning that pubs and breweries would be controlled by the local government. Joseph Chamberlain, MP of Birmingham, went even further. He had studied the Gothenburg system used in Sweden where citizen-owned companies ran the drink trade like a public utility. In 1877 he presented the Birmingham Town Council with a drink map showing drink dens in red to help him argue his point. This map does not survive, but it was widely written about at the time. With the Gothenburg system, he argued, 1,000 of the city's 1,800 beer shops could be closed, there would be no licences to deal with because shops would be run by town employees and owned by the town, and any financial benefit went directly to the town to fund schools, libraries and other improvements that benefited all residents. He believed that drunkards would become moderate drinkers under this self-policed system. His goal was not to completely forbid drinking, but to get a better handle on the worst of it. People did push back though – not all drinkers were receptive to government interference, while at the other extreme were those who did not want to tolerate any drinking at all. His community-owned licensing idea did not pass. In another town, an 1872 Act restricting closing times for public houses sparked riots and demonstrations. Even Prime Minister Gladstone believed his 1874 defeat was due to his support for restrictions on gin and beer. Other proposed reforms included county licensing committees and higher licensing fees. They all went nowhere, but the number of creative solutions reflect the recognition of there being a serious problem with alcohol consumption.

PARLIAMENTARY PRESSURE

In the 1864 session of Parliament, Lawson introduced his 'Permissive Bill' for the first time. The word 'permissive' might sound like it means to allow something, but here it meant to allow the *prohibition* of alcohol. The bill, leveraging the 'needs of the neighbourhood' objection to liquor licences, was designed to give a mechanism to the people who actually lived in the proximity of a public house that was applying for a licence to inform the magistrates whether the neighbourhood needed one. Under this model it would be possible to propose that no public houses were needed at all. A majority of local people would have had to express a desire for a licence for a 'need' to be established. The concept of 'need' had been amorphous since its introduction in the late 1700s. There had been many attempts to measure need, including distance (miles from the centre of town, paces, yards and minutes walking distance between licensed premises) and ratios (such as number of licensed premises per population). [5] The bill did not pass, and Lawson almost did not even get a second reading (required for a proposed bill to continue its passage to becoming a law) until the MP from Manchester spoke up. Lawson proposed the bill at nearly every session for the duration of his service, but it never got all the way through

– though margins did become narrower over time. Regardless, the Permissive Bill became Lawson's legacy. You can imagine him shaking his head at his colleagues when he wrote:

> The Public House was looked upon as about as sacred as the Church, and the idea of doing anything which might eliminate it from our national and social life was looked upon with horror. The very men who, from their position and from the accident of property, were able to prevent drink-licenses being obtained near their own dwelling-places could not think of allowing their poorer neighbours to have the same chance of protecting themselves.[6]

Variations of his proposal came to be known as the 'local veto' and sometimes the 'local option'. It could be argued that Lawson's long-view strategy worked though. In the next several years, micro-restrictions did pass, such as the Licensing Act (prohibiting the sale of spirits to children under sixteen among other things) and the Coal Mines Regulation Act (prohibiting minors from being paid their wages in drink shops) both in 1872, the closing of public houses on Sundays in Wales in 1881 and the Payment of Wages Prohibition Act (outlawing the use of licensed premises as the location for the payment of any workers' wages) in 1883.

The Members of Parliament in favour of temperance faced strong opposition from those MPs directly linked to and benefiting from the drink trade, such as the MP for Derby and owner of the largest brewery in England, Michael Thomas Bass. Even those MPs who were not directly benefiting from the alcohol industry still had constituents in the drink trade, including maltsters, farmers, publicans, hop growers, glassmakers and general working people who had grown accustomed to having beer nearby since the 1830 Act. There may not have been much to fear from citizens closing their local pubs even if the Permissive Bill had passed. The public wanted their beer.

Parliament was just one legal front the UKA tried to persuade by influencing who was elected and which laws might pass. Another front was at the local level, and included getting special maps before the eyes of those who could allow or refuse liquor licences – the magistrates.

DRINK MAP POP-UPS

Anti-drinking groups used a variety of persuasive methods. Some wrote songs, prepared sermons (pre-packaged and delivered to churches), printed poems, gave speeches, passed out pamphlets, put up posters, orchestrated large crowds to witness the signing of pledges to refrain from drinking, while others distributed maps. The most dramatic of those maps used sharp contrasting colours for impact.

By the late 1870s the use of persuasive maps was not yet part of an organized approach. But shortly after Dr Nichols posted his half-mile of London map in the *Alliance News*, the term 'drink map' first appeared on maps in three UKA auxiliary member townships. The first was the previously mentioned drink map of Birmingham of 1877, with articles appearing verbatim in at least four newspapers around England on the same day. The map itself has not survived and future drink maps of Birmingham did not display the words 'drink map' anywhere on their surfaces as a title or otherwise. This earliest Birmingham drink map used just one colour – red – to highlight drink spots, while later versions used many colours. The next two maps, circulated in 1878, which both used the words 'drink map' directly on them, were the *Drink Map of Southampton* (FIG. 11) and the *Drink Map of Norwich* (discussed in more detail in Chapter 5). Both of these maps also contained explanations on the reverse.

The *Drink Map of Southampton* was published by the St Mary's Church of England Temperance Society, which hoped to press their views on local residents and its Member of Parliament. Its message was dramatic:

11 This early drink map of Southampton has letterpress printing on the back with a message encouraging the closure of public houses. *Drink Map of Southampton*, 1878.

SOUTHAMPTON COMMON

Highfield   Highfield Ch   *Portswood*

P O R T S W O O D

RACE COURSE

RACE COURSE

SOUTHAMPTON CEMETERY

ALMA ROAD   AVENUE ROAD   SPRING ROAD

LONDON AND SOUTH WESTERN RAILWAY

★ BEER HOUSES
● LICENSED FOR BEER & SPIRITS
■ BREWERIES
▲ OTHER LICENSES

Nº OF LICENSED HOUSES 522

Newtown

CRICKET GROUND

Norham

LAND LET BY RECTOR OF ST MARY WITH COVENANT AGAINST PUBLIC HOUSES

WEST PARK   EAST PARK

ST PETERS Ch

COMMERCIAL ROAD

WEST MARLAND

Millbank

HOGLANDS

ST MARY'S Ch

ALL SAINTS Ch

DRINK MAP OF
SOUTHAMPTON.
1878.

PORTERS MEADOW

ST JAMES Ch

THE DOCKS

THE REASON WHY !! We were once reminded by Mr. Chamberlain, M.P., that Southampton is one of the most intemperate towns in England. The accompanying "Drink Map" is intended to show the inhabitants of Southampton THE REASON WHY. 'The struggle of the School, the Library, and the Church,' wrote the late Mr. Charles Buxton – one of the greatest brewers in the world – 'all united against the beershop and the gin palace, is but one development of the war between heaven and hell: were it not for intoxication, pauperism would be nearly extinguished in England, and crime of every kind would fall to one quarter of its present amount.'

The startling picture presented by this map suggests the following question: Is the law which gives those in no way representative of the ratepayers, the power to crowd the poorer districts of a town with these red spots, a just law, in a free country? Or should the ratepayers have the power of protecting themselves and their families from what is admitted on all hands to be the bane and curse of the land.

A brewer in favour of less drinking? Charles Buxton (1823–1871) was a partner in the brewery of Truman, Hanbury, Buxton, & Co. and a Member of Parliament, first for Newport, then Maidstone and finally East Surrey until his death. He had many interests – the education of children (he drafted and actively supported legislation encouraging compulsory education), birds, the social welfare of the industrial classes, architecture (he designed and built a fountain near Westminster Abbey in 1863 commemorating his father's anti-slavery efforts) and the effect of strong drink. In 1855 he published an article about it in the *North British Review* titled 'How to Stop Drunkenness'. In it he presents anecdotal evidence from judges saying that the majority of prisoners come before them on account of drink, from sea captains confirming that when there less spirit was allowed in sailors' rations there were lower incidents of trouble on board, from a reformer who spent a day in a Manchester prison and declared that every one of the 550 prisoners could trace their crimes to drink, another in Glasgow who determined that the paths in the wrong direction of 99 prisoners of every 100 jailed started with whisky. 'The alarming amount of madness in the United Kingdom is well known to be in great part owing to the abuse of fermented liquors.'[7] Buxton wasn't just a brewer – he truly believed that it was beer that could save society from the wretched transformative effect of drinking spirits. As an example, he compares gin at 50 per cent alcohol to a London porter at 4 per cent, and notes that it would be nearly impossible to drink enough porter to achieve the negative effect of a half pint of gin. (The use of a pint measurement when referring to a spirit itself flags the extent of the problem.) His proposals to counter spirit swilling were to cease the sale of intoxicants at 10 p.m., and to allow coffee houses to stay open without limit so people still had places to gather if they did not want to go home.

Drink maps were borrowed by all kinds of groups to argue various points. For example, St Mary's Church in Southampton, in addition to being home to the local temperance society, also hosted groups like the Southampton Cocoa Rooms Company that worked to address the drink problem. The 21 December 1878 edition of the *Hampshire Independent* ran minutes where shareholders considered whether its cocoa house should open on Sundays as a counter-attraction to public houses, specifically noting that the Drink Map of Southampton showed the town overwhelmed with 522 of them. One observer noticed that the Drink Map was also popular with ships' crews and sailors new to town. Another grumpy attendee argued that if Sunday was intended for worship then people shouldn't be socializing at the cocoa house any more than the public house.

MP brewer Buxton also endorsed the idea of the bona fide traveller exception to the hours that public houses were allowed to serve intoxicants. The bona fide traveller exception had already been a part of the licensing scheme for many years and was affirmed again in the Licensing Act of 1872. This was at a time when railway travel was

A VEXED QUESTION SETTLED!!!

The Licensed Victuallers' Chart
& Bonâ Fide Traveller's Guide.

COMPILED FROM THE ORDNANCE MAPS.

EXAMPLE—

A person starting from St. George's Church, Everton, on arriving at Tranmere Ferry, is a Bonâ Fide Traveller
A person starting from the Liverpool Town Hall, on arriving at Walton, is a Bona Fide Traveller.
A person starting from the Old Swan, on arriving at Kirkdale, is a Bona Fide Traveller

And Vice-versa.

Showing by the use of the Scale attached, the three mile distances from the Town Hall, &c., to any given point within the circles or squares, measurement of which are given.

Scale of one inch to a Statute Mile.

3 Miles

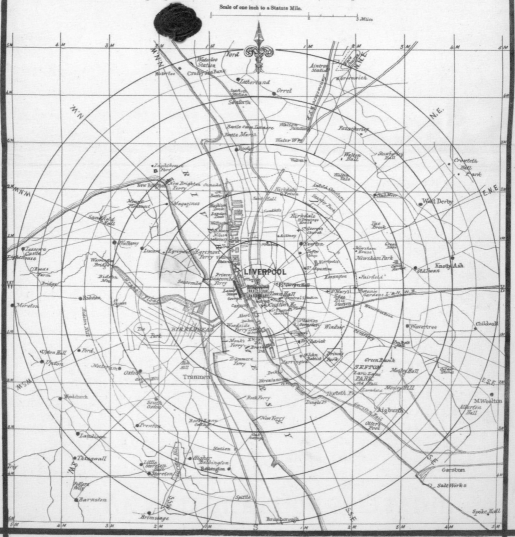

growing, but the horse was still the most common way to get around. The Act stated that a person could be supplied with liquor when a public house is otherwise closed if he has travelled at least 3 miles from where he lives, 'such distance to be calculated in a straight line on the ordnance map'.[8] The Act's rationale was that the physical effort of travel gave men a right to demand a beer in a public house after hours. Some public houses had a map available for just such circumstances, like the 1874 Ordnance Survey map of Liverpool titled *The Licensed Victualler's Chart & Bona Fide Traveller's Guide,* which had a string attached by waxed seal to a scaled card to easily measure 3 miles (FIG. 12). As one might imagine, this loophole was abused. The enforcement drove a wedge between the police and the publicans because when law enforcement raided a place it was the publicans who were fined more often than the drinkers, as the burden was on them to ensure that they were serving true travellers as defined by the Act. But other than asking how far they'd come, how were publicans supposed to know if someone was a bona fide traveller? Locals who walked or rode to the next town every weekend were technically within the law as long as it was 3 miles from where they lived. Buxton's idea was that the exception should be conditional upon the imbiber paying for a bed at the public house before he was allowed to order his first after-hours drink there. Amazingly the bone fide traveller exception stayed on the books well into the twentieth century.

Two more examples of drink maps prove how their makers hoped to use them to close drink shops. In 1880 the Gospel Temperance Union published a 'Drink Map of Stockport', describing it on its face as 'Giving a bird's-eye view of the chief cause of

12 This map of Liverpool was designed by John McGahey to help bona fide travellers measure, using a scaled card attached to the map by a cord and wax seal, if they had gone far enough to be allowed an after-hours beer. *The Licensed Victualler's Chart & Bona Fide Traveller's Guide,* 1874.

the misery, poverty, and crime in the town'. On the map itself is a sort of flirtatious wink to magistrates, carving out an extra-special message and affectionate pressure to take action: 'Dedicated to the thoughtful attention of all who wish the welfare of the Borough, and especially of the Licensing Bench of Magistrates'. The incomplete sentence grates indeed, but the intended audience is clear. The second map, possibly made by the UKA itself, is an 1882 'Drink Map of York', discussed in the 18 March 1882 *Alliance News*, showing 'in a single view a startling proof how sadly poor York is drink-shop-ridden and drink-cursed'. This particular map received a lot of criticism for inaccuracies, and this careful review of boundaries and 'fake' declarations became one way that the drink trade could attack and discredit the maps.

From Newton Heath, Aberdeen, York, London, Southampton, Stafford, Stockport and Liverpool these early drink maps were hundreds of miles apart. The concept of providing a stark image to highlight how many places there were to buy alcohol, and in which neighbourhoods, was popping up all over Britain. They differed in size and colour, in the messages appearing on them and who published them. Yet their popularity was spreading.

If a map were drawn to show the first drink maps, it would reach the far corners of the UK. But as a persuasive tool the splotchy drink maps had yet to become a trend. That is, until a teetotal London solicitor ignited the idea.

DETAILED INSTRUCTIONS

John Hayward was a humble solicitor generally working on cases of fraudulent horse auctions, mysterious deaths and debts for broken windows and stolen dogs. Yet he blew on the embers of drink-map activity when he outlined, with step-by-step instructions in a letter to the editor of the *Alliance News*, how to make the most convincing anti-licensing memorials (written or spoken statements summarizing a position) to local authorities shortly before the annual brewster (licensing) sessions in August of 1882:

> First, get a large-scale Ordnance map of the district, mark on it the site of the house for which a licence is sought, then stick a small black or red wafer or piece of paper on the site of the house already licensed, distinguishing the beerhouses and off-licensed houses by tickets of different shape. I have often seen the magistrates astonished by a map so decorated; in some districts it will appear literally pitted with public houses.
>
> Next get a petition respectably rather than numerously signed. The sight of your map would often assist you in getting signatures. Get first signatures from clergymen-ministers, church wardens, large employers of labour, doctors, lawyers, schoolmasters, and other representative men; then get as many signatures as possible from persons living in the immediate vicinity of the proposed public-house, and afterwards as many others as you can. The magistrates never read through or give much time to petitions. You should, therefore, have a synopsis ready, so that the solicitor whom you must instruct to present it may in a minute be able to inform the magistrates that it has been signed by so many clergy, so many ministers, &c, and so many persons residing near the proposed public-house.
>
> The prayer of your petition should be short: you need not enlarge on the evils of drink. I think the Alliance have models of the right sort, which I am sure they would willingly forward.

Hayward's letter went on to coach would-be licence resisters about what to expect in the way of reply and how to respond. The idea of a drink map as a persuasive propaganda tool was gaining momentum, but it wasn't until the decision in a notorious groundbreaking legal case dramatically shifted magisterial power that temperance groups galvanized their efforts around maps. That case was called *Over Darwen*, and its name was written on numerous drink maps after it was decided in 1882.

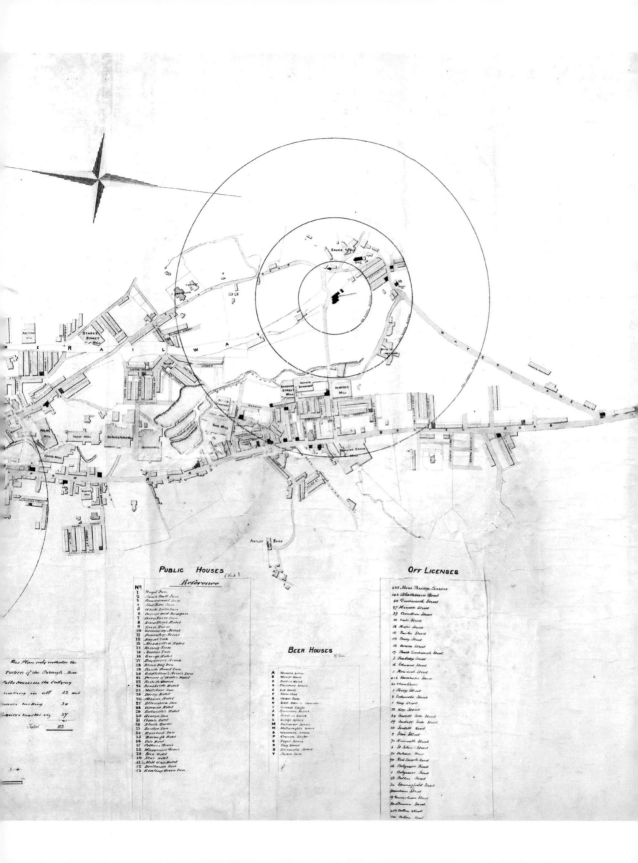

PUBLIC HOUSES (No.1)

Reference

OFF LICENSES

BEER HOUSES

Mo.

CHAPTER THREE

THE TEMPERANCE MOVEMENT LEVERAGES A LEGAL VICTORY

Picture a group of men studying a map showing all of the places to buy beer in a small town. The men are white property-owning crown-appointed magistrates in England, and the year is 1882. At this special licensing session they have to decide which businesses will have permission to continue selling beer. In previous years this was a mere formality, especially for places deemed protected, or grandfathered in, by the 1869 Wine and Beerhouse Act. Meanwhile public drunkenness was surging, as were poverty rates, incidents of crime, the number of abandoned children in rags – and the local magistrates had had enough. They meant to put a stop to it.

They planned carefully. A map, at their request, had been submitted as evidence for their consideration. It showed the location of every shop, pub and beerhouse. As the magistrates questioned the applicants,

13 (*previous spread*) This map was likely submitted as evidence for the magistrates to consult at the town of Over Darwen's Adjourned Brewster Sessions in 1882. Public houses are noted in red, beer houses in blue, and off-licences in green on the map and referenced in an incomplete column. Photograph of the *Plan of Over Darwen* (untitled), 1882. (*Location of original unknown; image courtesy of Roy Taylor of Darwen.*)

they consulted the map to assess the needs of the neighbourhood for each licence. After eight hours of discussion and study, they refused to grant half of the licences on the grounds that no one could complain because 'there was no part of the borough … more than two minutes' walk from some place where beer could be purchased'.[1] Their decision created a sensation and launched the widespread use of drink maps as tools of temperance persuasion. Here is the story.

A NICE GUY JUST TRYING TO SELL BEER

There is a town in the county of Lancashire nestled north-west of Manchester and south of Blackburn formerly called Over Darwen and today simply known as Darwen. It has a celebrated Jubilee Tower (that the locals call the Darwen Tower), a history of once-thriving cotton mills (and cotton riots) and a famous flash flood. Otherwise it's just another of the charming market towns dotting the English landscape. Or it would be, if a famous licensing case had not securely fastened it to the history of licensing law thanks to a grocer called Alexander Kay.

Alexander Kay's family had been 'Darreners' (as the locals of Over Darwen refered to themselves) for so many generations that Kay Street was already well established and still exists. Some branches of the family did not speak to each other, while others got on so well they married their cousins. Kay's shop at 334 Bolton Road likely carried pickles, sugar, tea and other sundries. This is important to note because Kay's business was already successful and not dependent on alcohol sales when he initially applied for an off-licence (meaning one to sell beer to take home, sometimes called a grocer's licence). He first applied for, and was granted, such a licence in 1875. He may have brewed beer in his shop, or he may have sold bottled beer; he did not sell cans, as that technology would not be perfected until the 1930s. Every year he applied for a licence to continue selling beer, and every year it was granted. There had not been any complaints against him by the local police or townspeople, and he was known to be an honest man and of good character. By 1882 his shop was selling between two and three barrels' worth of beer a week, which works out to be the equivalent of about 100 pints or four cases of bottles a day. His premises was valued at £800 a year when he applied for his licence to be renewed in 1882. That's nearly £100,000 in today's money. Alexander Kay, at forty-two years of age with a thriving business and a three-bedroom house with a fancy bay window, was doing well.

DISTANCE AND SCALE

The justices were looking at a map, yet their rationale was based on time – on how long it would take to reach another beer. A scale on a map is essential to help visualize distance, be it imperial miles or metric units. Yet when describing how far away something is, people tend to speak in terms of time and how long it takes to get somewhere more than measured physical space. 'Twenty minutes to the airport', 'three hours by train' and 'a fifteen-minute walk' are typical ways to explain how far it is to a destination regardless of technical distance. Ancient travel journals have recorded descriptions between starting and ending points in terms of a day's journey by mule or by camel. Therefore discussions by lawmakers studying a map of a familiar area that included how long it would take to walk between beers would not have seemed unusual. For example, when reporting on the town council meeting where a petition against licences was being debated, the *Blackburn Standard* of 11 September 1880 reported on the remarks of Reverend H. Moore.

> Within three or four minutes' walk of his house there were six or seven licensed houses. … He thought they might very well do with six licensed houses for the whole town, because he believed that this number could be so placed that there would be one within four minutes' walk of every house in the town.

In the end, a petition was adopted encouraging the magistrates against the granting of any new licences. He continued, 'if the magistrates would license more houses, such houses should be placed close to the gates of their residences-(laughter).' This was a direct finger-pointing at magistrates who granted licences in every neighbourhood but their own.

THE STATE OF THE LAW

In the same way that the Beerhouse Act of 1830 tried to correct for the gin laws of an earlier era, stop-gap laws were passed apace during the middle decades of the 1800s to try to regulate alcohol sales. One law would propose to restrict the hours that liquor-selling places could be open on Sundays; another would be passed in response with a list of exceptions. And so it went. Once Parliament passed the laws, the courts interpreted them – which added another layer of confusion. It became such an entangled mess that solicitors regularly wrote articles and even short books attempting to clarify the state of the law for other solicitors, magistrates and judges. Consistent ruling was a struggle. There was constant pressure on Parliament to overhaul the entire licensing system, but other matters took priority in Victorian times, such as expanding the Empire.

Whatever the laws were, magistrates were bound to implement them. And who were these magistrates? To qualify, if that's the right word, a potential magistrate had to be white, male and own land that provided a certain amount of income (thereby disqualifying the newly rich of the Industrial Revolution). They were appointed, not elected. And they were not required to have any legal training whatsoever. To paraphrase an observer of the time, magistrates knew more about dogs and horses than law and procedure. Fortunately they were guided by a clerk, whose humble title belied considerable expertise and rigorous qualifications as a seasoned solicitor to guide the magistrates on legal matters. Prior to 1877, to serve as a magistrates' clerk was considered a side role; an obligatory uncompensated duty imposed upon when called to the bar which was carried out with an unenthusiastic corresponding level of effort. That year an act was passed to professionalize the role, requiring – for the first time – that clerks be paid and have specific credentials. Other seemingly obvious new rules included forbidding the licensing sessions from being held in the local public house, sometimes the very place whose licence was being considered, as had been the norm in some towns.

What magistrates lacked in legal training they made up for in self-righteousness. Roughly a third of them were members of the clergy. This alliance sparked hope in temperance supporters that these quasi-judges cared about their local communities enough to prevent the opening of further drinking dens, even though the only consistent record magistrates shared across the country was preventing licences from being granted anywhere near their own homes. Decisions of the magistrates could be appealed to real legally trained judges, with the first stop being the quarterly Petty Sessions. If a party did not like that decision, had funds and a reason based in law, they could further appeal to the Queen's Bench. The last stop (for a few select matters) was the House of Lords.

Kay, like all other sellers of intoxicants, had to apply for a licence every year on a specific date set by statute. The event was called a Brewster Session. Every town had one, and they were usually held between the middle of August and early September. At these annual sessions, victuallers could apply to sell drinks to be enjoyed where people bought them as one does today in a pub or restaurant (known as an on-licence) or to take the alcohol home to drink (an off-licence, like Alexander Kay's). Licences were further distinguished by the type of alcohol sold, with different certificates for beer, spirits, wine and 'sweets' (English wine). There were even more permutations, at times up to fifteen different kinds of licence – and an entity could have more than one kind at a time. A place that was licensed to sell spirits, wine and beer was sometimes known as a 'full house'.

At the appointed time and place a brewster session usually began with a police report given as sworn testimony about the number of violations of any licences held by the applicants (such as serving after hours or watering down the spirits) and of drink-related crimes in the area (such as fighting or property destruction) to give the justices a sense of the level of drunk-and-disorderly behaviour compared to the prior year.

Next came the presentation of memorials, deputations and signed petitions. These could be from prominent individuals, but more often accompanied carefully worded assertions approved by town councils, temperance boards or churches expressing positions about licensing in general or opposing a specific licence application. This method of attempting to add weight to a given perspective through large numbers of supporters was common, especially when few people were permitted to vote. Declarations and petitions were accepted for hot topics of the day such as building permits, highways, dog licences and other things that were deemed in need of local control. The discussions of what to include in memorials were often heated. Their debate and approval was a solemn process and was taken very seriously. For women in particular, who were permitted to sign petitions, it was their only recorded means of influence. The memorials and declarations were not considered evidence the way a sworn police officer's statement was, but they were written into the minutes of the licensing sessions and a synopsis of the proceedings was often published in local newspapers, which allowed the information and the tone of the meetings to reach beyond the immediate audience. The magistrates were not required to listen to the statements or read them, and occasionally refused. But this was rare, as they understood it was an opportunity to give the public a platform even if they chose to ignore their positions.

Any objection to a licence being granted had to be made two weeks before the brewster sessions. In many towns it was to the objections of police that magistrates paid most mind. The drink trade was well aware of this; police records about drunkenness were notoriously unreliable in certain districts because of how many officers received courtesy refreshments from the public houses. Angry members of the public protested by calling out the uneven enforcement of licensing laws against individuals instead of public-house owners. The one exception to the two-week notice rule applied to the magistrates themselves. They could object at the start of the brewster session itself and then adjourn the session another two weeks while the applicants had time to respond and prepare. The public understood that the heading 'Adjourned Brewster Session' in newspapers meant that the meeting was for challenged licences. Unlike the routine sessions, the adjourned versions often drew crowds.

At the Over Darwen Brewster Sessions in August of 1882 the magistrates objected to every application to renew a licence to sell beer off-premises, all seventy-two of them. The notices, served by the local superintendent of police, Mr Bryning, declared that the licences were being opposed on the grounds of there being no necessity for them in the neighbourhood. In other words, there were too many places to get beer. Word spread through newspapers across the country and national attention turned to the tiny town.

THE 1882 OVER DARWEN BREWSTER SESSION

In an unusual move, when reporting on the results of the Adjourned Over Darwen Brewster Sessions, the *Blackburn Standard* of 16 September 1882 published the names and addresses of every challenged off-licence applicant, and included the number of yards from each licensed premises to the next. The list had been compiled as proof that a proper objection had been served on all licences up for renewal. Every licence holder in the crowded police court in the town hall where the session was held arrived ready. Mr Costeker appeared for most of the off-licence holders and Mr Broadbent appeared on behalf of the temperance groups and the public-house licence holders. Indeed, the interests of teetotallers and public houses were joined in this instance because the public houses saw the grocers as competition. They believed they were losing customers who could drink at home instead of going to the pub. This strange alliance is part of the reason why the drink trade struggled at times to present a united front. The anti-alcohol folks had their share of internal disagreements, but not to the extent of the open fractures within the drink trade.

There are indications that significant planning went into the magistrates' ruling well before this highly anticipated adjourned session met. For one, the municipality of Over Darwen was less than a year old, having just been constituted in September 1881. This means that the 1882 brewster session was its first, and the bench of magistrates was newly appointed and beholden to no businesses yet. It was also thirty-four-year-old solicitor Frederick Hindle's first time leading a group of magistrates as their clerk. On top of this, the state of licensing law was in flux, as usual. That year, just two weeks before the licensing session, a parliamentary bill referred to as the 'Off Licensing Act of 1882' (Beer Dealers' Retail Licenses Amendment Act, 1882) became the rule of law. Prior to the new Act, magistrates could only refuse to renew an existing licence if misconduct had been shown by the licence holder. (Another reason why the sessions began with a police report.) Hindle, as the clerk of the Over Darwen Magistrates, convinced them that they now had the power to refuse to renew licences based solely on whether the neighbourhood needed them. Does any neighbourhood *need* seventy-two shops that sell beer? It's curious that it was Hindle who steered the magistrates in this direction, since he had previously acted as the solicitor on behalf of some of the very same grocers when they applied for licences years earlier, although no potential conflict of interest was raised at the time. Three of the newly appointed magistrates had sat in Darwen and had actually ruled in favour of granting Kay's original licence nine years earlier.

Denying licences was nothing new. In previous years the magistrates of Darwen had denied plenty of applications for new licences. For example, in 1869 it denied sixty-seven licences. This was a special year, being the first time that gaining a licence had a higher bar than just paying the fee. Again in 1872 more licences were denied due to an active local tax-paying group that condemned the activities that tended to surround drunkenness. (Rate-payers, usually also having the right to vote, had more clout than the average citizen.) This trend of opposition was mentioned at this licensing session as well. Yet in 1875 Kay's licence had been granted. Perhaps off-licences were thought of as less threatening because they encouraged drinking at home in private. Anyone could support or object to an application, including local temperance groups, police who brought lists of violations or character complaints, and people living in the neighbourhood. Likewise, members of the drink trade could state their positions.

MAP EVIDENCE

The justices, at the suggestion of clerk Hindle, requested a map of all existing licences. This hand-drawn map, likely submitted as evidence and used by the Over Darwen magistrates to aid in their decision, has no title (FIG. 13). North is not at the top. It is not based on an Ordnance Survey. Its legend remains incomplete. Yet the name and location of existing public houses and beer houses are firmly planted in red and blue, labelled by numbers and letters. A third column contains a list of addresses which matches exactly in name and in order the list of off-premises licences granted renewals and published in the local *Blackburn Standard* the day after the decision. Each address corresponds precisely to the green marks on the map. Apparently there was no time or need to label them, and the map remains unfinished. Hand-drawn on the map are also circles showing the radius, in yards, to make it easier to see and count the density of licensed houses within them. The map has a hole at the centre of each circle, presumably where a pin and string or protractor point was stuck to draw even circles.

Frederick Hindle's account of the infamous brewster session reported it best in the book he published shortly thereafter titled *The Legal Status of Licensed Victuallers*:

> An elaborate plan of the district, showing the whole of the licensed houses, was put in evidence; an adjournment of the sessions was taken for the purpose of enabling the magistrates personally to inspect the whole of the houses objected to, and when, after spending nearly eight hours at the adjourned sessions in taking evidence upon the merits of each individual case, the magistrates refused the renewal of thirty-four out of seventy-two existing off-licenses, it was shown that they still left a larger number in proportion to the population than the average of most other boroughs in the county. … It was also shown that after the thirty-four licenses were struck off there was no part of the borough where there was any street or cluster of houses more than two minutes' walk from some place where beer could be purchased.[2]

The grounds for not renewing licences were announced and published in the *Blackburn Standard*. A few of the refusals were based on Licensing Act violations such as selling beer after hours, but most were based on there being no need for so many places for consumers to find alcohol. The question of 'need' drove much of the debate around licensing for the second half of the nineteenth century.

Twenty-five of the thirty-four rejected applicants appealed. When a decision is going to impact many similarly situated litigants, an economically sound strategy is to select the most likely set of facts to win as a test case to appeal everyone's matters. Depending on how that goes, the others can safely follow or live with the negative outcome. Here, Alexander Kay had the strongest facts. He was respected, had a clear record and had operated for years without complaint. It's likely that some of the rejected applicants did not file notices of appeal because of the expense. Legal fees were part of the gamble of the outcome. Only those who filed the appeal notices would benefit if Kay's outcome prevailed; the others lived with the refusal to renew their licences regardless of the outcome on appeal.

The *Blackburn Times* reported on the appeal on 16 December 1882, noting that 'Enormous interests depended upon it; it was not merely Alexander Kay, but the matter affected the whole of the off-licences of the country, and this case had been put in the fore-front.' The paper also noted that the struck licences were from three disreputable parts of town, including Bolton Road – where Kay's shop was located.

LEGAL TECHNICALITY

Kay's case was ruled against at the Quarter Sessions and again at the Queen's Bench, where the appeal stream ended and the ruling became the law of the land. Is an annual licence to sell alcohol a continuation of an existing licence, or a brand new licence? This is the deceptively simple question that became the centre of *Kay* v. *The Justices of Over Darwen*. The phrase 'legal technicality' is precisely what this case turned on. In Justice Stephen's ruling in favour of the magistrates he explained that the word *renewal* of a licence actually means to grant a *new* licence for the next twelve months. So *technically*, by law, each year when someone applied for a licence like the one they had the year before, they were requesting a new licence. A legal technicality, yes, with huge implications. Most importantly the holding meant that there was no such thing as a vested interest in a liquor licence. It had no lasting value and did not contain any equity that an establishment could bank on or borrow against.

Magistrates everywhere in England for the first time had complete discretion to refuse to renew annual licences to sell beer, meaning they finally had some small power to resist the juggernaut of the expanding drinks trade. Temperance supporters far beyond Over Darwen were thrilled and celebrating this victory. Probably with tea.

DRINK MAPS ARE BORN

Nine months after the November 1882 *Over Darwen* decision, the case name appeared for the first time on two simultaneously published drink maps, one of Liverpool and the other of Oxford. While they look quite different, the gist of the written messages fixed on the maps is the same: magistrates should use their newly confirmed power to refuse all licences to sell alcohol.

In spite of the victory of the case on appeal confirming that the magistrates had complete discretion to refuse to renew licences, the next year those same victorious Over Darwen magistrates buckled, and granted every single application to renew a licence to sell alcohol. They announced through their clerk that they had considered taking steps to further reduce the number of licences, but believed the government would transfer licensing power in the next year to a popularly elected body and therefore would not reduce the number without sufficient evidence that it was the desired outcome by the local population.

This is not actually surprising because the goal of the prior year had been to remove *unnecessary* houses. There wasn't a noted increase in population, so perhaps it would have been expected that all necessary renewals had already been granted. While the temperance groups may not have been happy, it could be argued that neutrality reigned and neither side had a gain or loss. Regardless of the Over Darwen magistrates' action the year after the controversial decision, the law was settled enough for temperance groups to finally have some sliver of legal force to reduce the number of licences, and they weren't going

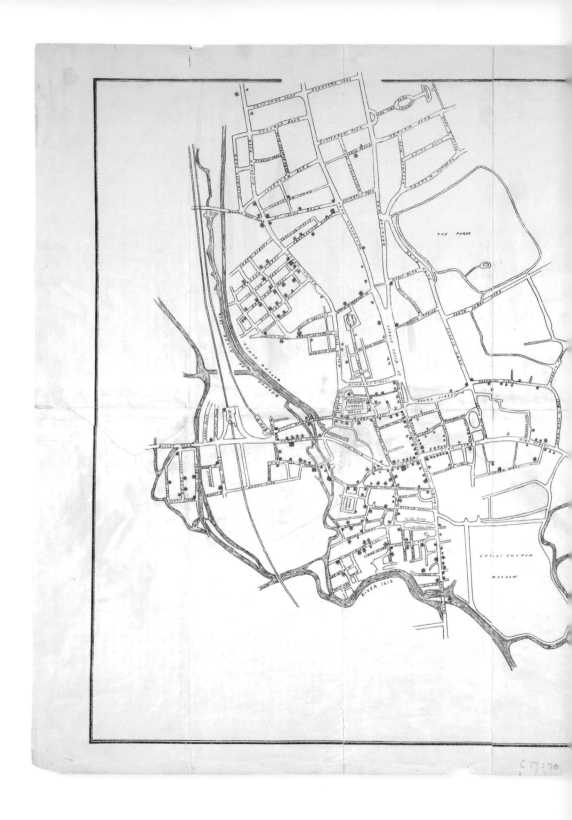

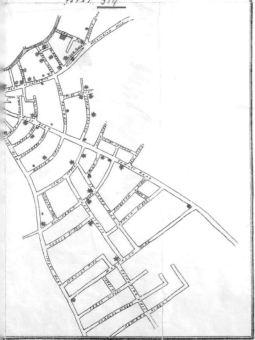

DRINK MAP

OF

OXFORD

1883

FULL LICENSED HOUSES --163
BEER HOUSES   135
BREWERIES   7
OTHER LICENSES   44
           TOTAL   314

THE DRINK TRAFFIC IN OXFORD.

THE "Drink Map" of Oxford makes bare the fact that few, if any, other towns in the kingdom are so liberally supplied with houses for the sale of intoxicants as this ancient city. Every twenty-second house in Oxford is a drink-shop, being at least fifty per cent. in excess of the average number throughout the country. The fair fame of Oxford is greatly tarnished by the effects of this over-supply of drink. Churches and chapels are plentiful, with ministers and Sunday-school teachers actively at work. Temperance and philanthropic societies are zealously striving to counteract the pernicious effects of drink. In spite of the efforts put forth by the Church and every other agency for good, there is an appalling amount of misery. Drunkenness abounds in our midst, and its attendant evils, crime and pauperism, are ever calling for attention. Can this be wondered at, seeing we have upwards of three hundred places licensed by law for the sale of strong drink?

Last year £126,251,359 were spent in strong drink, that is £3 11s. 7d. for every man, woman, and child, or £21 9s. 6d. for each family of six persons. The amount of money spent in Oxford in drink is not less than £140,000 per annum. One half of this is spent by the working-classes, who can least afford it. The result is idleness and ill-health, and very frequently poverty and crime. If one half of the money now spent in drink were spent in improving homes, it can easily be imagined how great a change there would be,—better food, respectable clothes, and comfortable furniture. Many working men might become their ... landlords, and even receive some rents to support them in old age. By this means a great impe ... which is now to a large extent paralyzed by the pauperising infl ...

That drinking, pov ... dy. Each pauper and criminal means in ... ortion of the community to pay. The ... re the makers and vendors of intoxicating ... he liquor-seller gets rich, several homes and ... o exist in our midst, the men engaged in it ... k-made paupers and criminals. It is an a ... community, but it is even a greater injusti ... ful effects.

All who look ca ... fact that drink-shops are not equally distri ... ey are most plentiful while in the north o ... rant a licence for any house near their ow ... of licences for houses amongst their poorer ... for the poor, but the gentlemen who have ... y refusing applications for permission to s ... eir friends do not wish public-houses near ... ions. But have not the

---

DRINK MAP

OF

OXFORD.

AUGUST, 1883.

Published by the Committee of the Oxfordshire Band of Hope and Temperance Union.

PRICE TWOPENCE.

Copies may be had from

J. M. SKINNER, Hon. Secy.,
10 Church Street,
OXFORD.

to squander it. A new power had been unleashed against the drinks trade, and the next step was to get the word out. What better way than on maps? Maps were quite useful when it came to grabbing the attention of impatient magistrates with a vivid picture to help them justify decisions, especially for those who could not be bothered to read the petitions of their own constituents.

TWO DRINK MAPS, ONE MESSAGE

The drink maps of Oxford (FIG. 14) and Liverpool (FIG. 15) are as different looking as the cities themselves. One is small and elegant, the other large and structured. The *Drink Map of Oxford* features quaint wobbly-rendered roads and borders, is hand-lettered, and tries so hard to focus attention on the drink spots that it leaves off the usual information one would expect on a map of this famous university city – including any reference to the university itself. Only the striking blood-red circles – each meant to be a vile reference to the devil's juice – draw the eye. Meanwhile, Liverpool's base map (meaning an underlying map, such as the Ordnance Survey Map that solicitor Hayward recommended in his 'how to make anti-licensing material' letter to the editor in Chapter 2) was initially a city plan with schools, churches, hospitals, railway stations and businesses. The ink of the base map is a brilliant royal blue, and the original list of places to explore takes up so much room on the map's face that its new anti-drinking purpose is not even mentioned anywhere on the front of the map. Its awkward title doesn't help either: *Mawdsley's Map of the City of Liverpool and Suburbs, 1883 Based on the Ordnance Survey and specially revised and corrected to the present time by William Wrennall, Surveyor*. It is only when turned over that the intention of the mass of scarlet red covering the front is revealed.

14 (*previous spread*) A framed version of this map was presented to the local magistrates for ease of consultation in hopes they would use it to deny licences to sell alcohol. *Drink Map of Oxford*, 1883 (front and reverse).

A second Drink Map of Liverpool published in 1891 is the same as the earlier edition on its face, with the addition of a red stamp on the front upper left corner with the title 'Drink Map' and a bare-bones legend. In dramatic contrast to the earlier map that had more words on the back than any other, on this one the back is blank. Apparently the crimson dots spoke for themselves.

Newspapers ran stories comparing the maps when they were published. Under a 'Temperance News' feature, the *North Wilts Herald* of 25 January 1884 reported on a meeting where the bright markings on the maps were analogized to sickness, commenting that Oxford appeared to be a city with the measles and Liverpool as more of 'a place where fever was prevalent (laughter)'.

Liverpool was a city with an unruly reputation. Licences were freely given. Compounding the widespread drunkenness was racism against the Irish population, corruption among police, densely packed dock workers, and a lack of cooperation between the owners of industry and the landed gentry. The city's 'grant them all' approach to licensing had notorious results. For an extended period it had the highest licensed-houses-per-resident ratio in Britain. Oxford had more of a town-versus-gown divide, although the various colleges' breweries did not require licences and are not shown on the map.

The maps had different temperance publishers, yet this distinction hides the tight network of anti-drinking groups responsible for them. Nathaniel Smyth compiled the Liverpool map on behalf of the clunkily

15 (*overleaf*) The face of this map does not explain what the blanket of red dots signify. Only on the back is there a second title, *Map of the City of Liverpool, with the licensed public-houses, beer-shops, grocers, confectioners, and other licenses, marked thereon.* A brief legend indicates that a red dot represents 'Full-Licensed Houses (1,898)' and a red *X* represents all other licenses (425), followed by eight panels of small-print stats, law and arguments about the evils of drink and how they should be addressed. *Mawdsley's Map of the City of Liverpool and Suburbs, 1883. Based on the Ordnance Survey and specially revised and corrected to the present time by William Wrennall, Surveyor, 1883.*

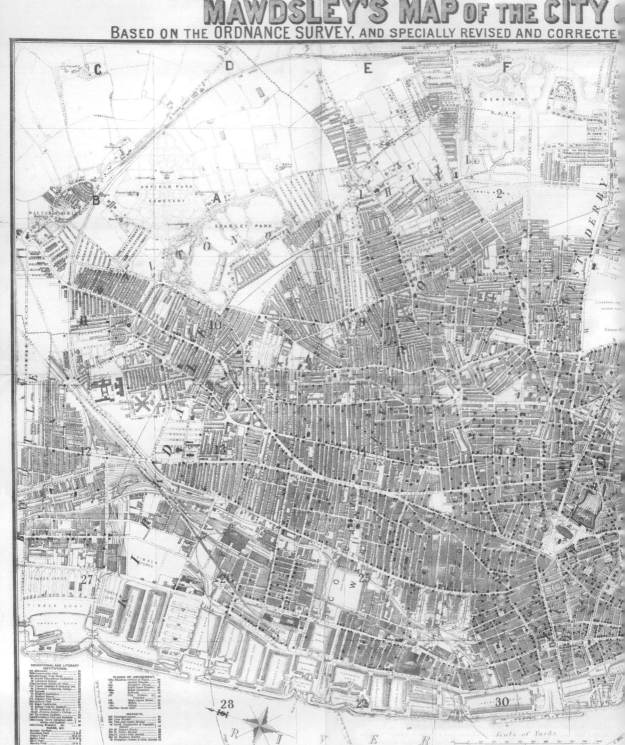

MAWDSLEY'S MAP OF THE CITY

BASED ON THE ORDNANCE SURVEY, AND SPECIALLY REVISED AND CORRECTED

CRPOOL AND SUBURBS, 1883.

ESENT TIME BY WILLIAM WRENNALL, SURVEYOR, &c. LIVERPOOL.

TOXTETH PARK CEMETERY

SEFTON PARK

Cricket Ground

Review Ground

PRINCES PARK

PROPOSED DOCK

MERSEY

NOTE.—The Plan is divided into squares simply for the purpose of indicating reference to any particular Building or Locality given in the following List. The numbers on the left side refer to objects or places, whilst the numbers and letters on the right side indicate the square on which they will be found. The circular lines indicate distances from the Town Hall.

CHURCHES

CEMETERIES.

BAPTIST.

UNITARIAN.

PRIMITIVE METHODIST.

WESLEYAN METHODIST.

INDEPENDENT.

METHODIST FREE CHURCH.

WELSH CHAPEL.

MISCELLANEOUS CHAPELS.

CHARITABLE INSTITUTIONS.

PRESBYTERIAN CHURCHES.

ROMAN CATHOLIC.

METHODIST NEW CONNEXION.

MONUMENTS.

PUBLIC BUILDINGS.

CONCERT HALLS, &c.

RAILWAY STATIONS.

PUBLIC BUILDINGS.

REFERENCES.

named Popular Control and Sunday Closing Laws Association. He held officer positions in at least two other anti-alcohol groups and supported many others. A month before a major temperance conference in Liverpool in 1883, he published an article in the *Alliance News* about how drink maps could be used to 'defeat the fearsome breweries that magistrates are afraid of'. At the conference itself, Smyth led a field trip to the docks to show up close how condensed the public houses were, and afterwards gave copies of his drink map to attendees. Many United Kingdom Alliance members attended the conference from England and Wales. Compare that to the drink map of Oxford published by Martin James ('M.J.') Skinner as an officer of the Committee of the Oxfordshire Band of Hope and Temperance Union. While he presented it on behalf of the Band of Hope, he also presented memorials and petitions on behalf of other organizations at various towns' brewster sessions, most often as the face of the United Kingdom Alliance. The Band of Hope advocated complete abstinence from alcohol with a focus on educating children. It organized sessions teaching about the evils of alcohol that ended with signing pledges promising to abstain from alcohol in front of witnesses, followed by choral entertainment (see facing page).

The maps, for all their differences, have one thing in common: strongly worded letterpress on the back citing the freshly decided *Kay* v. *The Justices of Over Darwen* case and directions about how the decision could be used to fight drinking. Liverpool's map has the most text of any extant drink map. It goes into considerable detail about the state of the law of licensing and the role, power and duty of magistrates on matters of granting, renewing, removing and transferring licences. It argues that magistrates have been manipulated by the drink trade and instructs them how to resist such pressure for the good of the public:

16 Band of Hope Pledge and Jubilee temperance ribbon, 1897. The ribbon would have been worn by temperance advocates to announce their allegiance and commitment to sobriety.

The Band of Hope and Temperance Union was founded in Leeds in 1847 by a Baptist minister after a child in his congregation died from alcohol. He became intensely motivated to prevent further deaths. The society's founding purpose was to teach children about the ills of alcohol, provide alternative activities as they grew older, and encourage future supporters to the cause. Their most commonly used device was called the 'pledge', a statement on a card that young people would publicly sign promising to abstain from alcohol. Blue ribbons were worn by some temperance supporters to announce their sympathy for being alcohol free. The use of ribbons as visible allegiance to teetotalism was also embraced by the US-based Women's Christian Temperance Union, which used white metallic bows. In the Indian army, special medal-style ribbons called Chamberlains were given out when an abstainer reached six months sober, with additional decorations added for successive periods. In the same way that temperance groups leveraged the spots on drink maps as a metaphor for sickness, they also borrowed the language and symbols of military battles in their fight against the drink trade. The Norfolk & Norwich Gospel Temperance and Blue Ribbon Union took its name from the ribbons. The Band of Hope in particular featured blue ribbons around the text of their posters, pledge cards and banners for special celebrations.

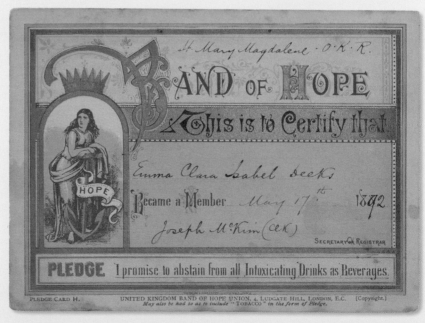

In the Over Darwen case, in the Court of Queen's Bench, November 24th, 1882, … Mr. Justice FIELD said – 'In every case in every year there is a new license granted. You may call it renewal if you like, but that does not make the license an old one. The Legislature does not call it a renewal, it calls it "a grant by way of renewal". The Legislature is not capable of calling a new thing an old one. The Legislature recognizes *no vested right at all in any holder of a license.* It does not treat the interest as a vested one in any way.'

In single-spaced text with slim margins, the back of the map packs in views on the drink topics of the day. A considerable portion of the map is devoted to the transfer of licences which were being used to circumvent hostile magistrates, beginning by saying how complicated the rules are, and then continues to make sure with its overly murky explanation that nothing is clarified. One section of the long commentary brings up the opportunities provided by multiple side doors in public houses to escape police detection ('It would be well if the back and side doors of the drinking Places in Liverpool were closed, as in Glasgow.')

Glasgow indeed had been using a drink map of its city to convince magistrates to prevent additional licences as early as 1859 when the usefulness of a drink map was brought up at a water-fountain support meeting. Glasgow, like Liverpool, had significant dock frontage with visiting sailors whooping it up in the city. The efforts to curb drinking were often also about the overflow of people that came, drank and quickly left – leaving their tawdry tarnish on the local area. A beautiful example of a Glasgow drink map was compiled and prepared by William. M. Oatts and published by Bartholomew stationers in Edinburgh in 1884 titled the *New Plan of Glasgow with Suburbs, Showing the Distribution of Public Houses, Licensed Grocers, Churches, and Branches of the 'G.U.Y.M.C.A.'* (Glasgow United Young Men's Christian Association) (FIG. 17). This map is especially interesting because of its base map, which was initially created by the local branch of the YMCA. The base map itself has a large intentional blank space for use by other

17 (*above and overleaf*) The base map offers the support of the Glasgow United Young Men's Christian Association in showing the distribution of public houses, licensed grocers, churches and branches of the GUYMCA. *New Plan of Glasgow with Suburbs*, 1884 National Library of Scotland, Acc. 10222/PR/6, folio 107.

entities needing a finished map of the city with room to promote their missions, and the YMCA locations would always be a part of whatever map resulted. It was much less expensive to purchase a base map than to have a map made from scratch. This allowed the YMCA to reach beyond where they might otherwise be able to in their own networks. An article in the *Glasgow Herald* on 27 May 1884 described a much larger version of a Glasgow drink map hanging behind the platform of assembled magistrates at City Hall and clearly visible to the angry public at a meeting about recent licence renewals. They were angry because, although the local licensing justices at the Licensing Burgh Court had refused to renew many of the licences, believing that one public house for every sixty-nine families was far too many, they were all overturned on appeal and granted. To this group, the haughty appellate judges simply backhandedly swiped the carefully deliberated decisions as if dismissing fumbling children, which was sometimes

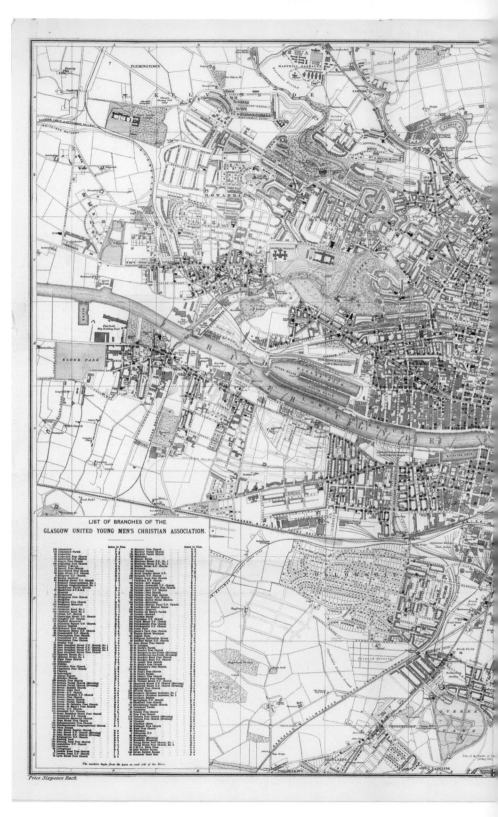

LIST OF BRANCHES OF THE
GLASGOW UNITED YOUNG MEN'S CHRISTIAN ASSOCIATION.

The numbers begin from the east on each side of the River.

how magistrates were thought of, especially in rural areas. But in a city the size of Glasgow the magistrates were solicitors, political figures and business leaders. It did not go down well. A resolution was passed calling out the justices for arbitrary rulings and exceeding their authority, and sent to the Home Secretary, who may or may not have read it.

At the conclusion of a typical temperance convention, well-known participants were mentioned by name in newspaper articles, as were the various temperance organizations which had a significant number of representatives attending. For example, a convention in London in 1881 welcomed eight UKA auxiliaries and over fifty different organizations from all over the UK, some designated by where they were from and others by specific topics of their club, such as a 'Local Option Union of Brighton' and the 'Scottish Permissive Bill and Temperance Association'. Because the names of individual participants were regularly published, it is known that the makers of the drink map of Liverpool and the drink map of Oxford had been at the same events, and likely knew each other. While neither map was made by the UKA on its face, both men – Smyth and Skinner – were heavily involved in that organization. This is important because the Liverpool map contains four lines of exact language used on another drink map distributed around the same time by the United Kingdom Alliance (see the drink map of Sheffield in Chapter 4). It is noteworthy that both maps' publishers embraced the UKA's emphasis on attacking drinking through the legal system, although both maps also include appeals on moral grounds. The temperance organizations were united in at least some of their positions and strategies.

PRESENTATION AND DISTRIBUTION

Once published, Smyth and Skinner broadcast the messages of their maps in creative ways. Smyth advocated to other ratepayers of Liverpool (who therefore had standing to oppose licences in court) to use the map as evidence to challenge licences; he claimed to have successfully done so himself. He also suggested approaching magistrates individually and personally, pulling out the map tucked under an arm to make the arguments directly, face to face, instead of waiting for a licensing session.

The Band of Hope's method for using a drink map became the most common, presenting them to the magistrates at brewster sessions. At the Oxford Brewster Sessions of 1883, the first session to decide on local licences in that city since the *Over Darwen* decision, Mr Hackney, a vice president of the Oxfordshire Band of Hope, presented the magistrates with a framed and glazed version of the *Drink Map of Oxford*. According to the *Oxford Times Weekly Supplement* of 1 September 1883, he handed it over with undisguised purpose. 'We present your Worships with a copy of a "drink map" of Oxford, which shows the number of licensed houses and their proximity to each other. We trust it may be of some service to you in refusing all new applications, and in considerably reducing the excessive number now existing.' It was accompanied by a written letter to the magistrates that closely, but not exactly, mirrored the language on the back of the map (which could not be seen when framed). The specific map presented at this session is not known to exist today, although available information supports that it is a copy of the same Drink Map of Oxford as is preserved in the Bodleian Library collections.

The minutes of the 1883 Oxford Brewster Session reveal a hesitancy on the part of the magistrates to deny any licence renewals without evidence of bad behaviour on the part of the applicants, as had been required before the 1882 Act. Frederick Hindle, the passionate clerk who led the Darwen magistrates, explained to the Oxford magistrates that the *Over Darwen* decision meant they could decline to renew as many licences as they believed were in excess of the city's needs. He encouraged the magistrates to object to all of the licences, as he had led the *Over Darwen* magistrates to do, and then review the drink map to carefully weigh the needs of the neighbourhood and then refuse all those in excess at the adjourned session. One magistrate confessed he had not heard about this new power. He was happy to have it, but went on to ask about the ability to appeal because they were afraid of litigation. Perhaps Hindle should have said that the magistrates could be sued if they *didn't* follow the ruling because they now had a *duty* to the neighbourhood to reduce unnecessary liquor outlets, but this did not occur to him. Magistrates, not being lawyers and therefore not staying on top of legal news, either didn't know about the case or remained confused. Of course Hindle couldn't promise that they would not be sued; he could only emphasize that he believed they would prevail. Magistrates shied away from any risk of humiliation because of the embarrassment of having their wrists slapped in public when their rulings were challenged. When magistrates tried to deny licences, the drink trade would look for an avenue around the ruling and, if they prevailed, would loudly publicize that the magistrate was wrong and against the interests of the working people. In spite of Mr Hackney's hope and Mr Hindle's assurances about their new powers, the justices denied only one application, adjourned one off-licence application and granted the rest of the licences that day.

The coverage of the brewster session proceedings by the local newspaper probably did more to spread the word to the public and other towns' magistrates than the letterpress on the map itself. The temperance movement was becoming savvy about influencing and educating through multiple channels at once. However, not everyone accepted the accuracy of the map. One annoyed reader complained in a letter to the editor that the *Drink Map of Oxford* was definitely inaccurate. He was sure of it because he could not find his three favourite pubs.

THE MAP MAKERS

What distinguishes these first two maps promoting the *Over Darwen* decision more than just what meets the eye is the men who made them. Both Smyth and Skinner held officer-level appointments in more than one temperance group. As officers, they appeared at brewster sessions to present memorials and petitions to the magistrates on behalf of their organizations and in the cause of temperance.

Smyth's primary allegiance was to the city of Liverpool. He was involved in the UKA, but only to the extent that their work had an impact in his home city. The Oxford publisher, M.J. Skinner, had a national agenda and was to become much more famous than Smyth in their lifetimes. By day, Skinner worked for the railway as an engineer. His passion was temperance, and he lectured on behalf of the cause all over the country. He made appearances, gave speeches and spread the temperance word from Bristol to Carlisle; from Cardiff to Harwich. Skinner rose up within the UKA hierarchy to be named District Superintendent for their south-east region, having moved from Oxford to Deal during his tenure. He was considered an upstanding citizen with a wife and three children. His star first began to fade when he was caught on a train trying to pass off his ticket from the same day the year before to avoid paying the fare. While that matter was pending in court, one evening Skinner made his usual rounds collecting subscription fees from several organizations that paid them to the UKA – and then he disappeared. His clothes were found on a beach between his last dues-collection site and the railway station he was expected to depart from. Articles lamented his likely death by drowning and there was even an obituary. One UKA branch held a memorial service for him. A few months later he turned up alive, well and on the run in the United States.

OVER DARWEN'S LEGACY

Thanks to the *Over Darwen* case, maps showing where drinks could be purchased were increasingly used to persuade magistrates and the public about the dominance of intoxicants in their midst and also as evidence in legal proceedings. The decision cemented the regular circulation of drink maps by towns all over England and also in Wales as an organized and coordinated strategic tool by temperance groups.

The attack was two-pronged: the visual impact of striking red dots on a map followed by inflammatory text. The strategy would be further fine-tuned and template language distributed among the United Kingdom Alliance's auxiliary branches over the years, and other temperance organizations increased their use of the method as well.

Drink maps, specifically titled as such on maps of Norwich, Southampton and York, already existed before the *Over Darwen* case. Similarly intentioned maps were published in at least two other towns in 1883: Newcastle upon Tyne and Derby. The Newcastle map did not survive, yet its use can be verified by newspaper accounts. The map was presented by Henry Scholefield on behalf of the Newcastle Auxiliary of the United Kingdom Alliance to the Board of Guardians. It was published at Scholefield's own expense. It was mounted, glazed and its spray of scarlet markings available for consultation in their boardroom, as were additional copies to hand out at the Central Station, Town Hall and police court. One onlooker wrote that if all public houses in Newcastle were joined it would make a 60-mile-long street.

The Derby drink map is particularly intriguing given that Michael Thomas Bass (1799–1884) was both owner of Bass Brewery and served as Derby's Member of Parliament. He took over management of his family's brewery in 1827 and grew it exponentially. His red triangle symbol, which looks a lot like the ones used to show 'licensed victuallers' or 'beerhouses' on many drink maps – was the UK's first registered trademark. He regularly voted in the interests of the drink trade, although he also gave money for projects considered to be distractions from the public house, such as parks and drinking fountains. Bass is owned by AB InBev today. The Derby map has bold black marks to show the number and location of licensed houses selling drinks, and uses five different shapes, instead of colours, to denote types of licence. Distances are instantly readable because of two prominent circles radiating from the town's population centres. The first is a 200-yard radius (encircling thirty-three licensed houses and one club) and a second is 220 yards out (encircling forty-nine licensed houses and two clubs).

The Derby drink map was prepared by Councillor John Wills, architect and surveyor, and funded by himself personally. As an architect, he designed a number of chapels in Derby. He presented the map to the Derby magistrates, with the support of Rev. G. Hunsworth, and also distributed 5,000 copies of the 1883 version entitled *Map of the Boro' of Derby shewing the number and position of Houses Licensed for the Sale of Intoxicating Drinks* (FIG. 19). Even though Bass was the MP for the area, articles emphasized that the magistrates had a stronger hand thanks to the *Over Darwen* case. A sceptical writer of the *Derby Daily Telegraph* reporting on the annual Brewster Sessions of 1883 on 31 August noted that 'The teetotallers appeared with the inevitable memorial, and they were further fortified by the "drink-map" of Mr. Wills.' Studying the yardage circles filled with marked drinking destinations, the same frustrated reader quipped that the map should be used to relocate licences from the 'overtavernised districts' to provide for the wants of the suburbs.

Derby's second extant drink map was also funded by Wills, in 1897. Both surviving editions are mounted on board, although the circulated copies would not have been. The later map is a bit larger, and is difficult to compare the two because Derby's city boundaries had changed and the radiating yardage circles were of a different number of yards. The later map shows a higher number of licences – 574 compared to 541 on the earlier map. This might seem to suggest that drink maps were not working, although no one mentions this at the time. Wills certainly could have made his point more clearly: while the overall population of Derby is shown on each map, the ratio of licences per person is not. It is only when he is interviewed that it becomes clear that in 1883 there was one licence for every 158 residents, reduced to one per 180 residents by 1897. One could read his account, or one could do the calculations – but why not just put it on the map? The later map also includes a chart showing

18 Bass secured the first two trademarks in the UK in 1876. The first was their famous red triangle, and the second was the diamond on this bottle. Advertisement from the Bass brewery, *c.* 1880.

19 Even the home county of the owner of Bass Brewery circulated a drink map. *Map of the Boro' of Derby shewing the number and position of Houses Licensed for the Sale of Intoxicating Drinks*, 1883. Derby Local Studies and Family History Library, 86 Cabinet 10.

MAP OF THE BORO' OF DERBY

SHEWING THE NUMBER AND POSITION OF

HOUSES LICENSED FOR THE SALE OF

INTOXICATING DRINKS.

JULY 1ST 1883.

ESTIMATED POPULATION 85574.

NUMBER OF ALEHOUSES		250
" BEERHOUSES [ON]		85
" " [OFF]		179
" WINE AND SWEETS		27
	TOTAL	541.

THERE ARE ALSO UNLICENSED HOUSES
WHERE INTOXICATING DRINKS ARE SOLD,
AS FOLLOWS

POLITICAL CLUBS		3
TOWN "		1
COUNTY "		1
	TOTAL	5.

KEY.

ALEHOUSES	ARE INDICATED THUS	"	▮
BEERHOUSES [ON]	"	"	▲
DO. [OFF]	"	"	●
WINE	"	"	▮
SWEETS	"	"	▮
CLUBS	UNLICENSED	"	✚

THE CIRCLE OF 280 YARDS RADIUS CONTAINS 55 LICENSED HOUSES AND 1 CLUB WHERE INTOXICATING DRINKS ARE SOLD

ST HELEN'S ST.
BRIDGE GATE.
KING ST.
ST MICHAEL'S LANE.
WALKER LANE
QUEEN STREET
FULL STREET
COLLEGE PLACE.
AMEN ALLEY.
IRONGATE.
250 YARDS
ST MARY'S GATE.
SADLER GATE.

THE CIRCLE OF 220 YARDS RADIUS CONTAINS 48 LICENSED HOUSES AND 2 CLUBS WHERE INTOXICATING DRINKS ARE SOLD

ST JAMES ST.
MARKET
VICTORIA ST.
ALBERT ST.
MORLEDGE
GREEN LANE
ST PETER'S ST.
220 YARDS
GREEN HILL

SOLD BY WILKINS AND ELLIS, ST PETER'S ST., DERBY. PRICE 6D
F CARTER, IRONGATE, DERBY.

JOHN WILLS, ARCHITECT AND SURVEYOR, MARKET HEAD, DERBY.

the convictions of individuals for drunkenness (6,800) compared to convictions of licensed victuallers (58) over the ten years 1887–1896. But he does not explain that temperance folks were angry that the laws were enforced disproportionately against citizens and perceived as protective of the drink hawkers. Wills – who was president of the Derby Working Men's Branch of the Anti-Vaccination League – also made the news when he was summoned in 1885 for not having his child vaccinated against cholera. He beat the charge when he proved that the notice was not properly served.

Another drink map meant to persuade local authorities was that of Cardiff, presented to a meeting of the Cardiff Social Reform Union in 1884 in support of their work. This group, like many at the time, took on issues of bettering the lives of its citizens through public-house distractions. It was prepared by Mr A.T. Davies, agent of the South Wales branch of the United Kingdom Alliance. The map has not survived in any local archives or libraries. There were a few who openly mocked the logic of showing where to buy alcohol as a means of discouraging drinking. In a letter to the editor shortly after this map was circulated, a reader lamented that there was not such a map available for the outskirts of Cardiff, so he might more easily find a pub nearby.

THE GROWING USE OF DRINK MAPS AS EVIDENCE

Submitting a map as evidence in a brewster session may have occurred before *Over Darwen*, but if so it didn't gain enough attention to be worthy of a newspaper account. Yet, after the technique was used successfully, the inclusion of maps as evidence of the location and density of licensed houses became so typical that it was anticipated in some places. Applicants began hiring surveyors and other experts to challenge any maps that might be introduced as inaccurate or misleading. Another tactic was to have a constable testify as to the distances between licensed houses – a kind of verbal map. While this was not as effective as the visual aid of a paper map, it did seem to carry some credibility as sworn testimony. The use of maps to persuade magistrates to at least stop granting new licences was getting attention. The *Daily Western Times* of 4 September 1883 quipped about the *Over Darwen* case with a nod to Charles Darwin, who had just died earlier that year: 'Over Darwenism, or the survival of the fittest in the "public" line, is a new phase of evolution.' Newspapers published articles noting and comparing several towns' drink maps, especially Oxford, Liverpool, Hastings and Derby. Even the Irish *Kilkenny Moderator* of 14 January 1885, wrote that 'The temperance societies are publishing various "Drink Maps" of various parts of England and Scotland.' But when Dublin was offered an Ordnance Survey map and information about how to add licensed public-house locations, the city politely declined. It's comforting to know that at least one city questioned the logic of showing a map of places to drink in order to discourage drinking. Or perhaps Guinness, already challenging Bass as the world's largest brewery, influenced such decisions in its backyard. It paid for a grand cathedral and for schools, so it had already mastered the art of appeasing the public.

As for Alexander Kay, he was so vexed by the outcome of the case against him and the loss of the ability to sell beer in his shop that he insisted on a special hearing to demand £50 for the use of his name when referring to the case. His claim went nowhere. A year later, the local newspaper listed Kay's home in a forced auction, suggesting he couldn't pay his mortgage, although there is no record of his shop closing, so it may be that his standard of living took a hit but his grocery business – which existed before he sold beer – continued. Clearly the sudden lack of beer sales impacted him economically. The court was trying to put public interests above those of individual business owners, yet it is by design in high-profile legal disputes for test cases to spotlight sympathetic characters. Poor Mr Kay.

ASSEMBLY ROOMS,

BELL STREET, HENLEY ON THAMES.

TUESDAY EVENING, THE ROYAL **NOV. 28TH, 1871.**

CHORUSES.
SONGS. DUETS.
RECITATIONS.

TUNE PLAYING.
MERRY PEALS.
CAMPANOLOGY.

WITH THEIR SPLENDID PEAL OF

FIFTY SWEETLY TONED BELLS.

POLAND STREET TEMPERANCE

HAND-BELL RINGERS

CONDUCTOR D.S.MILLER

H.HAVART. R.HOPKINS.

C.J.HAVART. W.SKINGSLEY.

AS THEY APPEARED
BY ROYAL COMMAND AT

OSBORNE HOUSE

BEFORE

HER MAJESTY THE QUEEN
And the Royal Family.

From Major-General Sir T. M. Biddulph, K.C.B., Keeper of Her Majesty's Privy Purse:

"Osborne, April 15th, 1870.—The Poland Street Temperance Hand-Bell Ringers had the honour of performing before the Queen and Royal Family yesterday, when their performances gave great satisfaction." T. M. BIDDULPH.

WILL GIVE THEIR POPULAR, TEMPERANCE, MUSICAL, RECITATIVE, AND

CAMPANOLOGICAL ENTERTAINMENT,

As given by them with marked success at Crystal Palace; Royal Polytechnic; Queen's Concert Rooms, &c., &c.

Under the auspices of the Henley Temperance Society.

ADMISSION:

Fronts Seats, ONE SHILLING. Second Seats, SIXPENCE. Tickets to admit 6 to Front Seats, FIVE SHILLINGS.

Doors open at 7.45.   To Commence at 8.15. p.m.

FOR FURTHER PARTICULARS SEE SMALL BILLS.

CHAPTER FOUR

THE DRINK MAP BOOM

While it was a sad day for Mr Kay when the *Over Darwen* decision came down, it was a turning point for temperance supporters. After chipping away on several barriers to their agendas to try to break through, they finally had an opening, and they meant to drive a stampede through it.

DRINK MAPS TAKE OFF

Fortified with a judicial blessing allowing magistrates to essentially close public houses if they decided they weren't needed, drink-map production exploded. Every town's temperance group wanted a drink map of its own. At least eighty-five different drink maps were published in the next several years. Only a handful are known to survive today.

On 11 August 1883, one year after John Hayward first provided step-by-step instructions on how to make a drink map suitable for convincing magistrates to refuse to grant licences in the *Alliance News* (see Chapter 2), he again provided instructions on how to make a

20 Temperance groups tried to lure public-house patrons to their alternative activities and soft drinks. Advertisements for Temperance Hand-Bell Ringers (1874) and (*overleaf*) Non-Intoxicating Beer (1892) – alcohol-free beer and wine have been around for a long time.

drink map for the coming brewster sessions. He strengthened his case by including the *Over Darwen* result and describing how it enhanced the position against drinking. This time, after advising readers to buy an Ordnance Survey map and mark in red all the places to purchase intoxicating drinks, he emboldened them further:

> It must be remembered that by the Beer Dealers' Licences Amendment Act, 1882, the licensing justices have full discretion over beer retail licenses both old and new, and at Over Darwen, in September last, 34 out of 72 existing off licenses were refused on the ground that they were not required in the district. The Court of Queen's Bench, in upholding the decision, has laid it down that there is no vested interest in a license, that the requirements of the neighbourhood is the most important consideration which ought to influence magistrates in deciding as to the granting or withholding a license. Temperance reformers are therefore now specially encouraged to aid the magistrates by their memorials to weed out the existing beerhouses which were established in days when they were beyond magisterial control.

Many of these maps were produced in large quantities and distributed to influence and educate magistrates, mayors, teachers, clergy, town councils, judges and others believed to be influential.

Of the surviving drink maps post *Darwen*, most mention the *Over Darwen* case by name. The strategic publishing of drink maps was motivated by the potential of improved political footing against the drink trade, and they were circulated all over England from as far north as Newcastle and as far south as Eastbourne, as far west as Plymouth and as far east as Norwich and sent to Members of Parliament. Two beautiful examples of maps created to convince citizens and politicians alike of the nefarious nature of alcohol are the drink maps of Sheffield and Birmingham.

THE DRINK MAP OF SHEFFIELD

Sheffield was slow to warm to the idea of using a drink map to send a temperance message, although the Sheffield Temperance Society had been active since 1831. James Silk Buckingham, Sheffield's MP, made an early effort in 1835 to form a committee in Parliament in an attempt to address the drink problem, but it didn't take off. Years later the infamous American teetotaller Neal Dow stopped in Sheffield on his 1862 lecture tour to share his strategy of resisting drink through legal means – but he did not inspire the locals. Things eventually started to percolate in 1873 when prominent local councillor William J. Clegg supported the taking over of a lapsed-licensed public house known as the Stag Inn to launch an alcohol-free gathering place as an alternative to pubs. He renamed it Star Home, and he dubbed it a 'public-house without beer' to much local fanfare. The Star Home offered billiards and other games, rooms for congregating, inexpensive meals, and… tea. A brewery called Truswell owned the building and leased it to the temperance group at nearly half the annual market rate, perhaps to claw back some goodwill after many failed attempts to secure licences for public houses tied to their beer. Opportunities to socialize outside of the public house – for example in cocoa houses, halls, hotels, temperance reading rooms and churches – were gaining ground.

In 1884 the Sheffield Brewster Sessions Committee (led by now Alderman William J. Clegg, the man who launched the beerless public house, was an officer of the Band of Hope and had long been involved in the United Kingdom Alliance) decided to circulate a drink map in Sheffield (FIG. 21). The map was first presented to a reportedly shocked audience attending a Church of England Temperance Mission meeting, and later used at subsequent brewster sessions to enlighten magistrates about the disproportionate concentration of public houses in poorer areas. Once they were dedicated to making a map, the designers took the potential of it as a visual medium one step further than other towns. -

The drink map of Sheffield is distinctive for three reasons: its language, its colourful graph and its title. First, some of the language on it is nearly identical to the previous year's drink map of Liverpool – suggesting that the circulation of drink maps was a coordinated persuasive tactic. The United Kingdom Alliance embraced drink-map production as a strategy and was attempting to engage in a cohesive national grassroots effort by providing members (including other temperance organizations) with template text for their local chapter publications consisting of sermons, petitions, letters… and maps. Compare the opening line on each map:

> *Liverpool* The accompanying Map (printed from the plates of Messrs. Mawdsley) is prepared in order that, at a glance, may be seen the number of Drinking Places existing in the City of Liverpool, and thus present an argument in favour of their reduction.

> *Sheffield* The accompanying Map (printed from the plates of Messrs. Pawson and Brailsford) is prepared in order that the large number of Drinking Places in our Town may be seen at a glance, that the *magnitude* of the drinking system may be more fully realised and thus present an argument in favour of their reduction.

Try to overlook the curious logic that glancing at a map full of pubs would be an argument in favour of their reduction. This is the only known example of nearly identical text on drink maps, although, given the UKA's well-documented historical use of model language in other mediums and the small number of drink maps known to have survived, there were likely others.

Just like the drink maps of Oxford and Liverpool, the Sheffield map quotes parts of the *Over Darwen* decision verbatim, as well as a statement made by the Home Secretary Sir William Harcourt supporting the decision:

> You have conferred on certain local authorities the right to say whether there shall be many public-houses or few, and I venture to say that it may be construed as the law, *whether there shall be any* in the locality in which they have jurisdiction, because the law is that every license is annual, and may be refused.

It also cites three different licensing acts. Drink maps became the space on which to disseminate the state of the law.

Next, the back of the map is divided into two parts. The left side is filled with letterpress in sections about the status of the law (including, by name, the *Over Darwen* case) and the expanded power of magistrates, a comparison of places to drink to the number of churches (10 to 1), and the detrimental financial cost of drink on everyone. Some optimistic calculations about the improved lives that would be had from the money diverted from drink are postulated – more bonnets and hats, boots, sacks of flour, teapots, cottages and town improvements.

(overleaf)

21 This map's title also serves as its legend. *Drink Map of the Town of Sheffield, shewing at a glance the number & situation of licensed houses of all kinds within the borough,* 1884.

22 This colourful graph takes up the right half of the back of the *Drink Map of the Town of Sheffield,* and lengthy text takes up the left side.

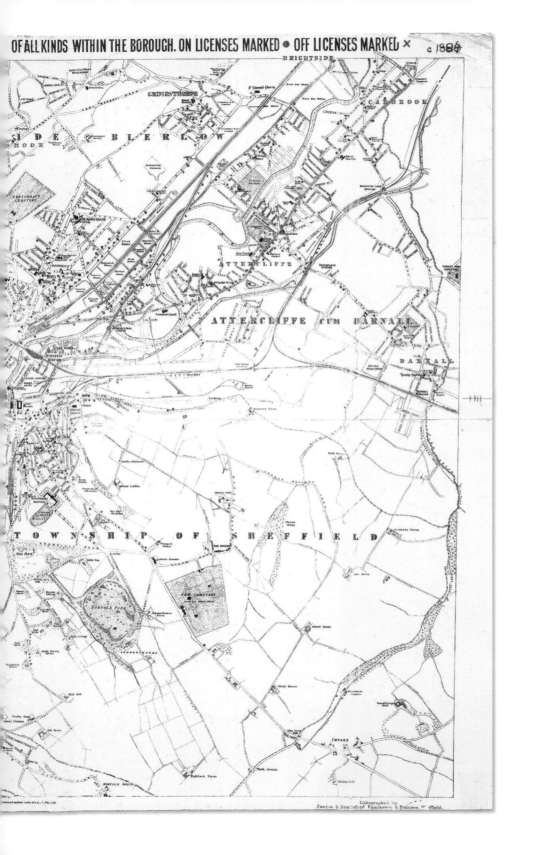

MAP

OF THE

TOWN OF SHEFFIEL[D]

WITH THE

LICENSED PUBLIC-HOUSES, BEER-SHO[PS]

GROCERS & OTHER LICENSES,

MARKED THEREON.

All Houses Licensed to Sell to be consumed on the premises are indica[ted]
thus ● (in Red).

All Off-Licenses indicated thus ✕ (in Red).

NOTE.—*There are in the Borough* 1,260 **ON** *and* 630 **OFF** *Licenses, making a* [total of]
1,890. *Some few of them are not here marked, because the Map does not take in t*[he]
localities where they are situated, the Borough being such an extensive one.

BREWSTER SESSIONS COMMITTEE:
TEMPERANCE HALL, TOWNHEAD STREET, SHEFFIELD.

1884.

DRINK MAP OF THE TOWN OF SHEFFIELD.

The accompanying Map (printed from the plates of Messrs. Pawson and Brailsford) is prepared in order that the large number of Drinking Places in our Town may be seen at a glance, that the *magnitude* of the drinking system may be more fully realised and thus present an argument in favour of (if not reduction.

The indications on the Map are placed as near as possible in the streets and roads where the licensed premises are situate, the same having been copied from the Police Register of Public-houses.

The Map has been prepared with great care, and may be taken as being approximately correct.

FULL AND UNRESERVED DISCRETION OF THE JUSTICES, &c.

It has been maintained for years past that the possessor of a Magistrates' Certificate and Excise License thereby secured a vested interest, and that the Magistrates were left without discretion in respect to the renewal of such Certificate. This view of the matter has evidently arisen through forgetfulness of the fact that the Act 9 Geo. IV. c. 61 has been and still is the basis of the Licensing powers, and that all subsequent Acts must be read and interpreted in the light of that measure.

The subject is one of such vast importance—not only to the persons directly interested in the Liquor Trade, but to the public generally—that it will be well to enquire whether we have legal justification for asserting that the Magistrates have full and absolute power to withhold Certificates "by way of renewal," if they should so decide.

In the first section of the 9 Geo. IV. c. 61 (the Act referred to), it reads, "And that it shall be lawful for the Justices * * * to grant licenses * * * to such persons as they the said Justices shall, in the execution of the powers herein contained, *and in the exercise of their discretion,* deem fit and proper."

MR. JUSTICE STEPHEN said—"The Legislature says, when we talk of a renewal of a license we do not mean that, but we mean a new license granted to a man who has had one before. It says, 'do not suppose that we meant a renewal, but the granting of a license by way of renewal.'"

Again, in the same case in the Court of Appeal, before Lords Justices, 12th December, 1882, the court below held "that *no license whatever existed longer than twelve months.*" It was the duty of the Magistrates yearly to revise such licenses. The Act of 1882 put 'beer off' licenses in the same position as 'full' licenses, which had no renewal as of right."

LORD CHIEF JUSTICE COLERIDGE was surprised to hear there was any right of appeal and would give no encouragement to such an appeal.

The Right Honourable Sir W. VERNON HARCOURT, M.P., Her Majesty's Secretary of State for the Home Department, said, in his place in the House of Commons, on April 27th, 1883,—" You have conferred on certain local authorities the right to say whether there shall be many public-houses or few, and I venture to say that it may be construed as the law, *whether there shall be any in the locality in which they have jurisdiction, because the law is that every license is annual, and may be refused.*

" Why should the Magistrates have power to *prohibit any sale* of an article of ordinary consumption? (Cries of 'No.') I say that *they have absolute power to do so.*"

Showing how wide-spread is the belief in the soundness of such decisions and declarations, the following may be read with interest :—The *Times* for February 19th, 1883, declares—" There cannot be a doubt that a misunderstanding has existed as to the position of Licensed Victuallers and other holders of Licenses. Rightly or wrongly there has grown up an opinion that somehow or other the discretion of the Justices had been limited by the licensing legislation of 1872 and 1874, but the sounder opinion seems to be that the discretion conferred by the Act of 1828 is unimpaired. It has hitherto been assumed that the holders of other than 'off' licenses have a kind of vested interest in them, but sooner or later we may find an effort made to induce Justices to refuse Licensed Victua[llers'] renewal of their licenses wherever they seem to be in excess of the requiremen[ts of the] public."

From the foregoing decisions of Judges and the statement of the Home Secr[etary it] will be seen that the Justices have *full* discretionary power to grant, transfer, ren[ew or] *refuse* to grant, transfer or *renew* licenses for the sale of intoxicating drinks.

NUMBER OF PLACES OF WORSHIP.

It may be interesting (if not humiliating) to note that whilst the Drink Sho[ps of the] Town total 1,890 we have but 196 places of worship—the proportion being nearly t[en to] one of the latter. If the late Charles Buxton, M.P. (Brewer), was right [when he] said "that the Beer-shop was fighting *against* the Church," is it not both amazi[ng and sad] that a Christian Legislature and a Christian people should be comparatively c[ontent to] allow the present deplorable state of things to continue ?

THE YEARLY DRINK BILL.

The Annual Expenditure upon intoxicating drinks in the United Kingdom has [for] many years quite appalling in its magnitude. From the Excise returns publish[ed some] months ago, it is shown that last year we spent (as a nation) more than 125,000,00[0 of £] upon these drinks, or three pounds ten shillings per head for every man, woman a[nd child] in the kingdom ; and be it remembered this enormous amount of money has b[een] expended in spite of all the splendid efforts that have recently been made by Gos[pel Tem-] perance Missions and the like to make the people *sober.*

SHEFFIELD'S SHARE IN THE EXPENDITURE UPON DRINK[.]

The people of Sheffield have for some years spent *annually* more than *a millio[n pounds]* upon intoxicants. The population of our borough is now about 300,000 ; its sha[re there-] fore, *last* year would be one million and fifty thousand pounds.

Suppose the money to be used in the following way :—

	£	
20,000 Pairs of Boots (Men's) at 10/ per pair	10,000	20,000 Dozens of Knives and Forks, at 5/ per doz
20,000 " (Women's)	10,000	20,000 Poor People each a Gift of £10 for a Visit to the Seaside, &c.
20,000 Hats, at 5/ each	5,000	
20,000 Bonnets, at 10/ each	10,000	5,000 Cottages Newly Furnished, at £40 each
20,000 Men's Suits of Clothes, at 60/ each	60,000	10,000 Cottages, New Furniture to replace Old, at £20
20,000 Women's Dresses, at 30/ each	30,000	Pay District Rate
20,000 Children Clothed, at £4 each	80,000	Leaving for Town Improvements
20,000 Sacks of Flour, at £2 each	40,000	
20,000 Pen Knives, at 2/6 each	2,500	Total
20,000 Metal Teapots, at 7/6 each	7,500	

If the money were spent as above, such a welcome impetus would be given to home industries that the cry of bad trade would almost cease to be heard in our midst ; trade now languishing would improve by leaps and bounds ; and, better still, hearts that are now weighed with drink-caused sorrow and despair would be made to leap and bound with joy and gladness.

In conclusion, is it not worth while to ask ourselves (speaking locally)—Can we reasonab[ly hope] the drinking in our own town to decrease and the people to become permanently sober whilst o[ur magis-] trates continue to license 1,890 persons to sell those drinks that make people drunk ?

The *Times* said, some three years ago, "that public houses mocked us, shamed us, confo[unded us] and baffled us at every point." Will they cease to do so if allowed to exist as now carri[ed on in] such appalling numbers as we find them in our midst to-day?

Sheffield Brewster Sessions Committee,

President, Alderman W. J. CL[...]
Treasurer, Mr. W. RICHARDS[...]
Hon. Sec., Mr. R. CLIFT HO[...]
Agent, Mr. W. H. HALL.

Offices—Temperance Hall, Townhead Street, Sheffield.

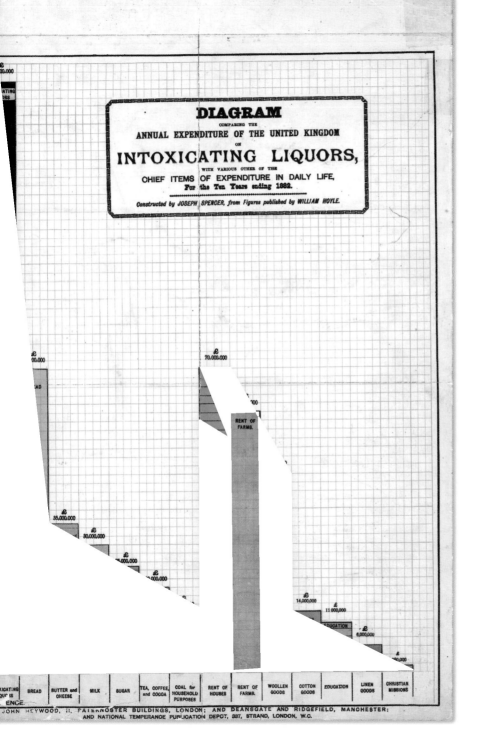

DIAGRAM

COMPARING THE

ANNUAL EXPENDITURE OF THE UNITED KINGDOM

ON

INTOXICATING LIQUORS,

WITH VARIOUS OTHER OF THE

CHIEF ITEMS OF EXPENDITURE IN DAILY LIFE,

For the Ten Years ending 1882.

Constructed by JOSEPH SPENCER, from Figures published by WILLIAM HOYLE.

£ 70,000,000

RENT OF FARMS.

£ 35,000,000

£ 30,000,000

£ 14,000,000

£ 11,000,000

EDUCATION

£ 6,000,000

INTOXICATING LIQUORS	BREAD	BUTTER and CHEESE	MILK	SUGAR	TEA, COFFEE, and COCOA	COAL for HOUSEHOLD PURPOSES	RENT OF HOUSES	RENT OF FARMS.	WOOLLEN GOODS	COTTON GOODS	EDUCATION	LINEN GOODS	CHRISTIAN MISSIONS

JOHN HEYWOOD, 11, PATERNOSTER BUILDINGS, LONDON; AND DEANSGATE AND RIDGEFIELD, MANCHESTER; AND NATIONAL TEMPERANCE PUBLICATION DEPOT, 337, STRAND, LONDON, W.C.

The £1,050,000 result: 'hearts that are now drooping with drink-caused sorrow and despair would be made to leap and bound with joy and gladness.'

On the right side of the back of the map is a dramatic graph with more colours than any other drink map (FIG. 22). Scottish statistician William Playfair introduced the bar chart in 1786 – just as the love of measuring and counting to lend an air of science to social problems was on the rise. The brightly coloured chart fills up the whole right side of the map's verso, bordered on the left with a menacing top-to-bottom bold black stripe proclaiming how much money was spent on alcohol. The other bars – shades of yellow, orange, pink, green and blue – represent the necessities of life with a visual emphasis on the comparatively smaller amounts being spent on housing, food and clothing to that of alcohol. Some disparities among the bars seem peculiar, such as the amount spent on bread being twice as much as shelter. This map has the Oxford map's elegant colour scheme and the Liverpool map's urban details – but its graph is unique to Sheffield.

The map was supported by the Band of Hope Temperance Union through subscription, meaning they paid as an organization for membership to the UKA and extra for the map's publication, and likely other temperance groups as well who were aligned with the political pressure and message they hoped to deliver. This pooling of resources and convergence of drink map messages is probably why the temperance lobby was stronger than the drink trades, at least for a while, as it was sometimes pitted against itself in fundamental ways that prevented a more collaborative defence or approach to the temperance voice's growing credibility and strength, as noted about public houses routinely appearing at brewster sessions to lodge objections to grocers' licences.

Finally, Sheffield's anti-drinking map presents the clearest message of all the drink maps included here because of its commanding and concise title in capital letters across the top, spanning the entire width of the map. It is the only drink map to use only two symbols on its face to denote the different types of licence, an *X* and an *O*, meaning one could either drink it there or take it home. Finally, the title is utilitarian because it doubles as the legend.

The 1884 drink map of Sheffield was still being used in 1887 when it was reportedly consulted at that year's local Brewster Sessions. The concept of how far one had to walk to get a beer raised by the *Over Darwen* justices was holding well, as a discussion among magistrates about the number of paces between licensed houses was noted at the session, as well as the number of houses and drink shops that were already licensed within 500 yards of each applicant's location. For example, on 31 August 1887 the *Sheffield Evening Telegraph* reported on an objection to a licence. Someone pointed out that, '[A]ccording to the "drink map" there were seven [public] houses within three hundred yards of the applicant's [public] house.' Licence denied. The snappy conversation back and forth between the charismatic William Clegg (here as a solicitor appearing for the interests of neighbours opposing licence applications) and Mr Binney (generally appearing on behalf of the drink trade) are listed in the minutes of those sessions with much laughter noted. The dynamic, jovial and self-deprecating Clegg is clearly the winner of many arguments. Take this exchange from the same day's Brewster Sessions report where it is discovered that a man has signed petitions both for and against a licence application. Mr Clegg finds it hard to believe there is any support at all for the application, and he challenges a witness to name even one person who could possibly back it:

Mr. Clegg What is his name?
Witness Wilson.
Mr. Clegg What number does he live at?
Witness 18.
Mr. Clegg And he has signed a petition against it. (*Laughter*)

Mr. Binney I have a Charles Wilson 18, Mount Street, who has signed my petition in favour of the application. (*Renewed laughter.*)
Mr. Clegg Oh, he is a gentleman who signs everything that comes before him evidently.

Such banter takes up a lot of column space over the many years that Clegg appeared at Sheffield's Brewster Sessions. On this same day, Clegg represented the interests of a neighbourhood fighting against a place called Fisherman's Rest that already had a beer licence from gaining an additional licence to sell spirits. Part of the testimony concerned how many paces someone would have to walk to find spirits elsewhere – naming three such establishments and complaining of how many steps were between them, specifically 460, 900 and 940 paces. (Try this for yourself – it's not very far.) The licence was refused. The newspaper accounts on these generally dull licensing meetings are consistently more detailed in Sheffield than any other city, no doubt in part because Clegg's wit made for such entertaining reading.

The Clegg family was full of lively characters, and they influenced much of Victorian-era Sheffield. The patriarch, William Johnson Clegg (advocate of the beerless public house mentioned earlier and founder of Clegg & Sons solicitors), lived until 1895. He was a member of several temperance organizations, including the British Temperance League, the United Kingdom Alliance, the Gospel Temperance Union and various local temperance groups. His son, eventually knighted Sir William Clegg (of Brewster Session entertainment) served on the Sheffield City Council for forty years and managed to lead many lives, including playing football for England and famously doing so in a match against Wales on the same day that he defended an accused murderer, Charles Peace. Peace was accused of killing the husband of a woman he had been stalking. The story of the crime, known as the Banner Cross murder, was so notorious that it was made into a play in London.

Drink maps were quickly outdated; usually foldable, portable and not made to last. Apparently their expendable nature meant they did not last in people's memories either. In 1903 – nearly twenty years after the original drink map of Sheffield first appeared – a visitor to Sheffield's Council Chamber was so amazed by a giant 6-foot-high by 12-foot-wide 'Drink Map of Sheffield' on exhibit there that he suggested in a letter to the *Sheffield Independent* that it be reproduced on a smaller scale and circulated. As if there had not been one before. This new wall-sized version had been created at the suggestion of the magistrates and made by the city surveyor's staff using information supplied by the chief constable. It is not known to have survived, but it was definitely different from the 1884 map because it had an updated number of drinking spots. Sheffield's earlier drink map showed 514 fully licensed houses, 623 beerhouse keepers, 598 off-beer licences and 75 other kinds of licence, making a total of 1,810 premises. Temperance pressure had been successful in Sheffield. In less than twenty years, the 1890 drink map showed that the total number of licensed houses had been reduced by more than 80.

BIRMINGHAM'S DRINK MAPS

The 1891 drink map of Birmingham is one of the most beautiful surviving drink maps. It is smartly colourful with burgundy, navy blue and olive green markings plus additional information to ponder on its surface. This city had at least two other maps, one made in 1877 (discussed in Chapter 2) and another in 1885; only the 1891 edition is available to examine (FIG. 23). The 1885 version was made by the local temperance organization and the city surveyor and presented to the Birmingham magistrates to aid their licensing decisions. Unlike the 1877 map that reportedly had only red markings, the 1885 version had red, blue and green dots like that of 1891 titled *Map of Licensed Houses, Birmingham*. This map is unlike most other drink maps in that it does

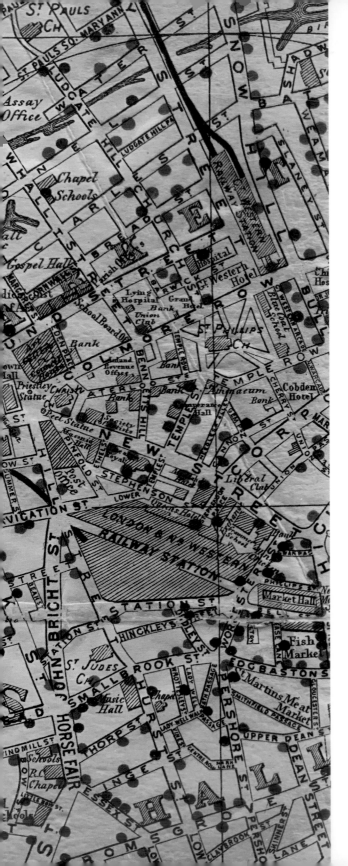

not categorize its markings by on- and off-premises consumption, but instead distinguishes among places that had full licences (victuallers) and just beer and wine licences. Red stood for licensed victuallers who sold both on and off premises (different from beer houses in that they could sell spirits as well), of which there were 653. Blue represented beer houses selling beer and wine both on and off premises, of which there were 1,026. Green spots were for 'off licences of various kinds' – 499, for a total of 2,178 options for intoxicating beverages.

Two further factors make the drink map of Birmingham stand out: its chart in the lower right corner and an isolated beerhouse in the centre of a workhouse property. The chart casts a spotlight on the vast difference between how harshly the drink laws had been enforced over the five years prior against individuals

23 Notice the number of intoxicating choices near the railway station. *Map of Licensed Houses, Birmingham*, 1891 (*detail*).

when compared to business owners, as the drink map of Derby also indicated. Without even reading the numbers, a glance reveals only double-digit violations against the drink trade, whereas cases against individual drinkers were in the thousands. The unwritten message was in the disparity – individuals would not be getting so drunk if the trade weren't oversupplying them, yet law enforcement punished only the drinkers. The second curiosity on this map is in the upper left corner. Near the eastern tip of the compass rose and across the river from the 'lunatic asylum' (where habitual drunkards were sometimes locked away) is a vast property with a large building on it labelled 'workhouse'. In theory a workhouse was a place where debts to society could be paid, literally, through work, and where the poor were fed and protected. In reality, as many social reformers who studied them in the day noted, they became the last resort for women-led families and men too old or too disabled to work. This map reveals a common situation: a drink outlet not just near the property, but directly on it and part of it. In fact there is no way to get to this beerhouse other than to be at the workhouse.

Beer was provided to workhouses by the counties where they were located. This rationing was a point of contention in some areas, to the point that workhouses got their own drink maps. In 1893, for example, the Workhouse Drink Reform League produced a choropleth map of England and Wales (meaning one with gradual shading) to show which counties had the highest expenditures on alcohol going to workhouses. The map was discussed all over England. The *Derbyshire Courier* on 7 January 1893 noticed that it was the southern counties that had the highest expenditure on alcohol for workhouses, while the *Leicester Journal* on 6 January 1893 pointed out that 'One thing that strikes the observer upon glancing at the map is the concentration of "alcoholic extravagance" in one part of England, broadly speaking, that in and around the Metropolis.' It was a reader in Somerset who seemed most outraged as he vented in the 18 March 1893 *Weston-super-Mare Gazette*: 'I cannot conceive why the paupers of Somerset and their caretakers should drink twice as much as those of Devon, and thirteen times as much as the men of Cornwall, and nearly twenty-five times as much as the Welshmen of Cardigan!'

As for the lovely 1891 Birmingham drink map, initially only two copies were made, and they were larger than this one. One was framed and presented to the Council House to hang in a committee room, and the other was for the police court where magistrates would also see it when brewster sessions were held there. After these two were published, the Birmingham auxiliary of the UKA arranged for 2,000 copies of it, identical but reduced in size, to be lithographed for distribution.

With so many maps featuring demonized drink dots, there was at least one made to encourage visiting public houses as attractions where beer could be enjoyed. It was the 1890 *Map of Cambridge shewing public houses and beer houses, houses with grocers' and wine licences and breweries.* This map is black and white with the various colleges and land around them in green, and red spots and squares to show where the beverage establishments were located. The big difference between this and temperance drink maps was the corresponding key listing the names of the forty-four drinking destinations that could be found on it – paid for by advertisers.

MORE LEGAL AID FOR TEMPERANCE: SHARP V. WAKEFIELD

It would be a stretch to pinpoint a singular purpose or strategy for the creation of drink maps before the *Over Darwen* case, but afterwards the focus became razor-sharp. Drink maps were directly aimed at magistrates. Their simple (and beautiful) images showing clusters of red on maps were efficient and more to the point than the long, dull written petitions that went ignored. This appeal to magistrates at the neighbourhood level was thought to be the best avenue for people to get their will across to those who could do something that would make an immediate difference to their lives and surroundings. The maps succinctly revealed to magistrates the extent of a problem that was otherwise invisible to them. They weren't known to ramble around the poorer parts of the very towns they served.

Many cities and towns charged a small amount for their drink maps to cover the printing costs or to raise money for temperance projects. The publisher of the Liverpool drink map, for example, noted in its annual report that while the run of maps cost £60, its sale helped to keep the organization out of debt the following year. The 1886 *Drink Map of Salisbury* published by the Salisbury Temperance Associates, which one observer said looked like the town had a disease, was sent all over England and Ireland with the goal of raising money to build a gym and other drink distractions, as was the 1885 *Drink Map of Bristol* published by one of the seventy-one regional Gospel Temperance Unions. Newspapers reported that several towns were using drink maps to further raise awareness and possibly funds. A map with the opposite approach of these – one published in 1886 to show all of the dry districts in the United Kingdom – was known as the *Temperance Map of England and Wales*. As it was the only one of its kind, apparently it did not catch on.

At the same time that drink maps spread across England instructing magistrates to use their new power to close existing beer dens, doubt was stirring. In truth, the law was not as well settled as the temperance groups had hoped. Reasonable people disagreed, in spite of the decision. A collection of cases known as the 'Barrow cases' upheld the opposite of *Over Darwen*. Notwithstanding Frederick Hindle's endeavour to explain the new legal muscle in his book clarifying the state of the law, the discretion given to magistrates was largely distrusted and ignored. It took another case in a neighbouring county and endorsement by the House of Lords to solidify the magistrates' authority. That case involved a public house called the Bridge Inn in the County of Westmorland owned by Susannah Sharp, a fifty-five-year-old widow. The Inn was on a remote country road and had been serving boozy beverages for thirty years when its annual licence was suddenly denied on the grounds that the neighbourhood did not need it. At some point in the proceedings the rural location of the inn became an issue, and Susannah Sharp's legal team actually agreed that the neighbourhood did not need a pub. They continued to insist that a public house having operated for so long with a clean record had a legal right to a renewed licence. They were wrong. The justices had a point to make, and the casual words about the neighbourhood not needing the pub were turned into a confession of sorts and used against Sharp to deny the licence on 20 March 1891. The House of Lords' decision was unanimous. The legal fees for each expensive appeal had been paid for by the Licensed Victuallers' Protection Society, which had a huge stake in the outcome, as many licences it had been treating as vested and valuable became worthless when they lost. The judgment confirmed that deals based on long-term liquor licence values would have to be renegotiated, and creative accounting and sketchy investment began to emerge in attempts to cover losses. To add to the drama, less than a month after the final judgment Susannah Sharp was found dead in her own well under suspicious circumstances. An investigation ruled her drowning a 'sad accident'. One wonders.

A second small burst of drink maps followed the judgment featuring news of the case printed directly on them. A 'Drink Map of Hastings' (1891), which unfortunately has not survived, was mentioned in the 18 September 1891 *Cork Constitution* as quoting part of the *Sharp* v. *Wakefield* case on the map itself, much like earlier drink maps did with the *Over Darwen* case:

The map showed effectively the grouping of licensed places, the great mass of the 245 of course 'planked down' in the working class regions, and so thick on the ground that not even Lord Bramwell could have thought any more to be necessary. The back of the map contains a mass of most useful letterpress, in the course of which many arguments are adduced to prove that in Hastings, as elsewhere, the number of licences is excessive; that the justices have power under Sharp v. Wakefield; that the working classes suffer most from the infliction of public-houses.

Additional drink maps that were circulated after the *Sharp* decision include St Albans (1891, presented at a temperance garden party in between a brass band and other speeches prepared by Mr McIlwraith, who lamented the lack of popular control over licences); St Leonards (1891), Bilston (1892), Norfolk/Kings Lynn (1892; this map survives with letterpress noted on the side by Foster & Bird, lithographers); Salford (1892, in the Salford Archives); Plymouth (1893, complained about by an annoyed local who claimed he could throw a stone from one licensed house to the next through the town); Chelsea (1893) and Chatham (1894). In Chatham the 25 August 1894 *Evesham Standard & West Midland Observer* ran this account of the recent Brewster Sessions:

> The temperance reformers, as represented by 14 Church, Nonconformist, and Roman Catholic Ministers, one Alderman, and four Town Councillors, who joined in memorial to licensing magistrates at Chatham on Tuesday to refuse to renew superfluous licenses, had armed themselves with a new weapon of offence. This took the form of a drink map, showing in red the situation of all the licensed houses in the Parliamentary borough.

In spite of the map and the argument, the magistrates renewed all of the existing licences. They did, however, refuse to grant any of the applications for new licences, so there were no public-house grand openings that year. In Wales a map was published in the 16 February 1894 edition of the *Brecon County Times*. The black-and-white square image took up two columns across, headed with the words 'DRINK MAP OF BRECON 1893' and below it – in cursive writing – an explanation that each dot represented a licensed premises, for a total of fifty-six drink licences. An accompanying article reports that magistrates owned 25 per cent of the licensed houses in the town. The United Kingdom Alliance was popular in Wales, so it is likely that there were more drink maps circulating. At least three drink maps were made in 1899: Sunderland (published by the United Temperance Society), Birkenhead (by the Birkenhead Vigilance Committee; this map is held at the Wirral Archives) and Maryport (referred to as the 'Drunk Map of Maryport' in a letter to the editor of the *West Cumberland Times*). Unless noted, no copies have been found of these maps. Yet.

JUDICIAL DISCRETION

How were the 'needs of the neighbourhood' to be determined? The magistrates were told to use their 'judicial discretion', but how is a magistrate's discretion different from a biased opinion? The concept of judicial discretion would be argued for years afterwards in licensing cases as well as many other types of legal dispute, and the concept is still litigated today. In an 1886 issue of the *Brewers' Guardian,* Sir Wilfrid Lawson, then serving as a magistrate, declared that he intended to vote down all licences. His words were used against him to show that he was not acting judiciously, that he was approaching the matter with a pre-formed decision. Two weeks later on 7 September in the same publication, a *Punch* portion included the following poem:

24 This map is annotated in pencil to show the new tally of licensed premises in 1904. There were ninety-six more than when the map was first published. *Drink Map of Leicester*, 1886, Leicestershire & District Temperance Union.

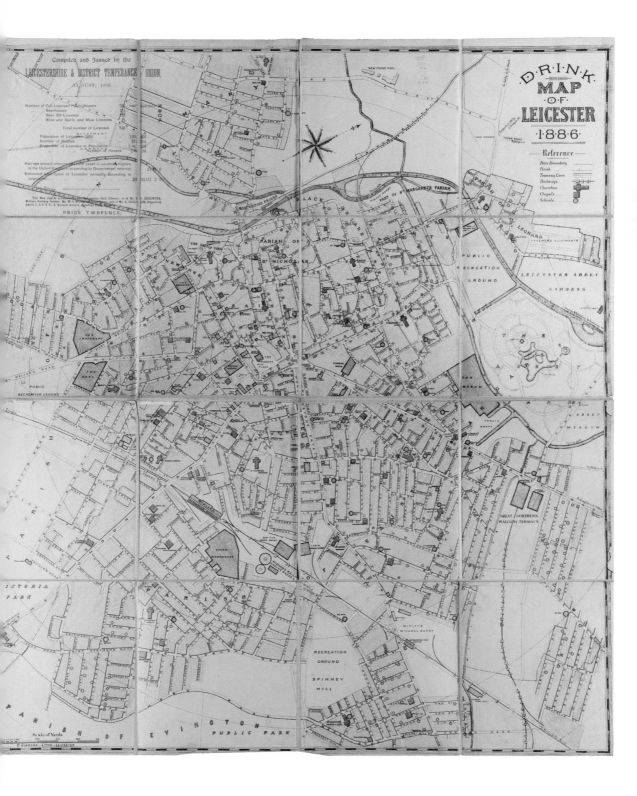

Sir Wilfrid Lawson, on the Bench says, 'Here,
I will not grant a Licence to sell beer.'
As magistrate his conduct is not flawless,
Let's change his title to 'Sir Wilfrid Lawless.'

DID DRINK MAPS WORK?

The production and use of drink maps peaked by the end of the
1800s. But did they actually work to convince magistrates to reduce
the number of places to buy a tipple? In at least one town a look at its
drink map showing annotations over time proves that the maps were
not always convincing. The 1886 Drink Map of Leicester (FIG. 24),
published by the Leicestershire & District Temperance Union, contains
information originally printed on the map tabulating 280 fully licensed
public houses, 174 beerhouses, 269 off-licences, and 25 wine and spirit
licences for a grand total of 748 alcohol options. Markings all over the
map added by hand illustrate that it was consulted and updated over
the years. A penciled year '1904' appears in the margin, with new tallies
for the licences in a column revealing a higher final total of 844 drink
locations. Most of the increases were in the form of take-away and wine
licences, while the number of beerhouses declined – which was consistent with national trends. Additional marks were made in pencil, in
blue pen and in yellow. New buildings were drawn onto the map, some
red spots indicating licences are crossed out with large blue V-shaped
slashes and occasionally the name of a pub is scribbled on. Leicester's
gravitation towards adding sipping venues rather than reducing them is
an outlier though, as the number of licences in the United Kingdom as
a whole declined by the end of the century.

25 The featured guest of this temperance celebration, George Cruikshank, was known for his
dramatic caricatures depicting the ravages of alcohol (see figures 29 & 30 for examples of his
drawings). Notice of a tea meeting held by the Leicester Temperance Society, 1850.

CHRISTMAS
TEMPERANCE FESTIVAL.

The Committee of the Leicester Temperance Society respectfully inform the Public that a

TEA MEETING

WILL BE HELD IN THE NEW HALL

ON FRIDAY, DEC. 27, 1850;

On which occasion the celebrated Artist & Caricaturist,

ESQ.,

Has kindly engaged to Preside, and Addresses will be delivered by

G. GREIG, Esq.

Of Leeds, formerly Professor of Elocution & Belles Lettres in the Andersonian University, Glasgow,

AND

MR. S. SMITHARD,

Late Temperance Missionary at Hull, Devonport & Ipswich, who will also sing a variety of

POPULAR TEMPERANCE MELODIES.

Tea will be on the Tables at Half-past Four o'clock—TICKETS for which, 1s. each, Children under twelve 6d., may be had at the Temperance Office, Belvoir-street; Cook's Temperance Hotel, Granby-street; and of Messrs. DONALDSON and FAIRS, the Society's Agents, at the Temperance Meetings. Admission after Tea—at about 7 o'clock—Twopence.

T. COOK, PRINTER, 28, GRANBY-STREET, LEICESTER.

NORWICH & NORFOLK
TEMPERANCE SOCIETY
ESTABLISHED 1836.

PATRONS.

The Right Honorable the EARL of ALBEMARLE.
BENJ. BOND CABBELL, Esq., Cromer Hall.
Rev. H. A. BARRETT, Chedgrave Rectory.

A TEMPERANCE

TEA MEETING

IN CONNEXION WITH THE ABOVE SOCIETY WILL BE HELD ON

TUESDAY, NOV. 6, 1860,

IN THE

LECTURE HALL,
ST. ANDREW'S.

After Tea the Chair will be taken by that Celebrated Artist

G. CRUIKSHANK, Esq

THE MEETING WILL BE ADDRESSED BY
THE RIGHT WORSHIPFUL THE MAYOR,

J. H. TILLETT, ESQ.

The Rev. J. Lee Warner, President of the Society,
Rev. T. B. Stephenson, and other Friends.

**TEA AT FIVE O'CLOCK PRECISELY. Tickets, 9d. each, Patrons'
Tickets, 1s. may be had of JARROLD & SONS, of the Members of
the Committee, and of the Ladies' Committee.**

**Mr. G. CUSHING the Temperance Missionary, will give a brief Report
of his Twelve months' labour in the City.**

*The Committee especially invite their fellow Citizens to attend this important meeting, and feel assured that the success already given will ensure
future aid from those who have, during the last 14 years, contributed to the support of a Temperance Missionary in the city.*

JARROLD AND SONS, PRINTERS, BOOKBINDERS, AND STATIONERS, LONDON STREET, NORWICH.

CHAPTER FIVE

DRINK MAPS IN MANCHESTER & NORWICH

Many towns and cities turned out more than one drink map, which makes it easier to track their persuasiveness over time. Birmingham, Derby, Liverpool, London, York and other cities had more than one version of a drink map. Two cities with multiple maps stand out: Manchester and Norwich.

Norwich had three drink maps of similar design, yet published by different groups. They trace a path spanning twenty-five years from 1878 to 1903. Manchester had a series of three related maps plus a fourth unassociated drink map – all created in one calendar year stretching from 1888 to 1889 – each with a specific purpose and meant for different audiences. To compare these cities' drink maps, let's start with those of Manchester.

DRINK MAPS OF MANCHESTER

Given that it was the location of the headquarters of the United Kingdom Alliance – arguably the instigator of the drink-map phenomenon – it is a bit surprising how long it took for Manchester to have a drink map of its own. By 1889 at least fifty United Kingdom Alliance auxiliary chapters had already published or endorsed drink maps of their towns. At a minimum, two of those maps used identical language printed directly on them to ensure the circulation of the UKA's unified message of change through legal authority instead of moral arm-twisting. It had been six years since the UKA's own newspaper, the *Alliance News* (printed in Manchester), had provided instructions on how to make a drink map no fewer than three different times. When the Manchester drink maps were finally made, they came out in quick succession.

This is not to say that the people of Manchester were complacent about drinking. As a city, Manchester competed with London for having the most organized, active and animated temperance population, and certainly won in terms of percentage if not actual numbers. The *Manchester Guardian* (today's *Guardian*) frequently published details of local brewster sessions and gave more column space than most to the summaries of those proceedings, including which licences were granted or denied. These lengthy articles often contained names of the magistrates hearing the applications, names of the applicants themselves, and whether they had ever been arrested or fined for alcohol-related infractions, the names and sometimes addresses of those seeking to sell alcohol, the arguments made, the outcomes and even commentary. If a teetotaller wanted to stalk a licence applicant, the *Manchester Guardian* provided the path.

It's possible that a drink map of the city of Manchester had not been prepared at the time of the *Over Darwen* decision simply because of the wide readership already relying on the *Manchester Guardian* for up-to-the-moment news on all things related to the proceedings. At the time of the decision there was considerable coverage describing the case and the impact it would have on the newly broadened authority of magistrates to clamp down on existing licences. Like the language displayed on drink maps of other towns, these articles gave people a plan of action to educate and influence their local magistrates. The message itself had already spread through this widely consumed medium; there may not have been a need for another outlet.

The history of the Drink Map of Manchester is easy to learn because it's written right on the front of the final version of three identical maps distributed in different ways. The lines explain that a first map was commissioned by the mayor, William Batty, to be given to the city's magistrates for easy consultation as they considered licensing applications. This map was created with a similar idea to the one submitted to the magistrates to mull over at the Over Darwen Brewster Sessions, although probably for more general and ongoing consultation. No copy is known to have survived. A second map was published on an entire full page of the *Manchester Guardian* on 5 January 1889, with a lengthy article titled 'Licensing in Manchester' describing the importance of the map, with the title on the facing page. The article took up nearly two columns. For a paper with only twelve pages, devoting this much space to a lone map must have been a stunning visual. Even in black and white, the dots, squares and triangles clustered in obvious areas would have provoked a reaction. The article opposite the map begins by noting that the six symbols 'honeycombing' the map actually represent fifteen different kinds of licence, such was the complexity and confusion of the law, and points to the correlation between the thickest portions of the map and the poorest neighbourhoods. It goes on, oddly, to relieve the magistrates of blame or responsibility because the law restricted their decisions. ('On this point some explanation will be found below, showing that, great as is the evil, it is due not to remissances on the part of the magistrates, but to the existing state of the law.') This was after the *Over Darwen* case but before *Sharp* v. *Wakefield*, and therefore confirms that the *Over Darwen* decision was widely ignored, misunderstood or disbelieved.

The feature continues, explaining that the 1869 Act governing beer houses gave those established before that year a sort of protective tenure. Ever since the Beerhouse Act, licences for beer houses were granted by collecting an excise tax. There was no duty on the part of the tax collector to consider the public good; they were simply to fetch the money. Licences were freely granted. The 1869 Act abolished excise as a licensing mechanism and moved all liquor-licensing authority to the local magistrates. It provided certain threshold requirements to apply for licences, such as good character and lack of alcohol-related convictions. This commentary is clunky, overly complicated and lacks a clear point. Which may be what led to the more elegant, concise and persuasive language printed on the face of the third and final map in this series.

The appearance in the newspaper of this second Drink Map of Manchester sparked a deluge of letters in response, both supporting and against it, for many days and even weeks afterwards. Some of the letters reacting to the map were sent to *other* newspapers, referencing the *Manchester Guardian* – an indication of how far the controversy reached. The intention was to use the growing popularity of newspapers and growth of literacy among the general public to spread the shocking image and keep the tension and urgency of the drink problem on people's minds. Remarking on the visual impact of the full-page map, with a nod to the growing use of electricity, one appreciative reader commented in a letter to the editor a couple of days later on 7 September that 'the "drink map" illuminates with almost electric effulgence what before was dark and unrecognised by the public.' Judging by the number and ardour of letters to editors that followed in the days after the map's publication, it worked. Under the long-time editorial eye of C.P. Scott, the circulation of the *Manchester Guardian* had grown more than ten times from its beginnings in 1852 to the end of the century. While it may be difficult today to understand what the big deal was with drinking at the time, it is telling that a newspaper anchored in social causes would give so much space to a temperance issue.

The third map of Manchester in this connected series is entirely red, both the roads outlined and the symbols representing the availability of drink, so its impact is muted compared to other maps (FIG. 27). The contextual and historical importance of the first two versions of the map is printed in black text next to the scarlet image. Unlike many drink maps that used the space on both sides to convey their messages, the back is blank; it was meant to be seen at eye-level on a wall instead of unfolded and turned over.

The written account on the map itself addressed three main areas: the visual impact of the clustered houses; the different types of licence, their legal status and how that influenced magistrates' ability to act; and who benefited the most and least when licences were transferred. Curiously, the *Over Darwen* case is not mentioned by name, although it is alluded to. By this point, there was widespread scepticism about the decision, and the *Sharp* v. *Wakefield* case was already in the pipeline.

Just over a month after the map's appearance in the *Manchester Guardian*, the 16 February 1889 *Alliance News* advertised copies of it:

> The striking 'drink chart,' which appeared recently in the *Manchester Guardian* has been reprinted by permission of the proprietors of that paper. It is now printed on very superior paper in ink of two colours. An explanation of the connection of the drink traffic with the present social condition of the people of Manchester accompanies the map. To insure a large circulation, single copies will be sent post free for 2*d*., and eight copies for 1*s*.

The most distilled and lasting language about the Manchester maps is printed on the face of the third version:

27 (*overleaf*) The United Kingdom Alliance published thousands of copies of this map after a full-page version printed in the *Manchester Guardian* received national attention. *The Drink Map of Manchester*, 1889. On the following pages the text of the map is reproduced in full.

THE DRINK MAP OF MANCHESTER.

▲ LICENSED VICTUALLERS

● BEER TO BE CONSUMED *ON THE PREMISES*

▣ " " " *OFF* "

L BEER AND WINE " *ON* "

O SWEETS " *ON* "

X OTHER LICENSES

THE DRINK MAP OF MAN

THE Drink Map of Manchester
duction of an instructive docume
for the information and use of th
trates. It appeared in the Mancl
of January 5th, 1889, and is now
permission. It shows at a glan
position of the liquor traffic in the g
the textile industries. There are
different kinds of licenses for the s
cants recognised by the law ; but fo
purposes the various branches c
traffic may be accurately indicated
the six distinctive signs employed i
panying map.

First, we may take the licensee
There are 484 hotels and public ho
are indicated on the map by triangu
The Justices have absolute power
application for a new license. Th
absolute judicial discretion to refuse
of a license previously granted.
done not only on the ground of the
of the license holder, but for any
satisfactory to the magistrates, so
consider each case judicially upo
There is a power of appeal to t
Quarter Sessions, which in this parti
has gained an unenviable notor
manner in which it has disregarded
wishes of neighbourhoods anxious t
from the curse which the liquor
into every district where it gai.
The applicant for a license may
point of law to the Court of Que
The powers of the Justices in reg
fers are in substance the same as
new grants and grants by way of re

Beerhouses are indicated on the
circular dot. There are 1,323 of
The Beer Act of 1832 is a mon
folly of legislators. It was passed,
warning and opposition, with the av
of promoting sobriety ? Give peop
opportunity to drink beer, it was sa
will cease to drink spirits. The
was given. On the mere paymen
the Excise any one could open a ho
they quickly swarmed in all great
caused an immense increase of drun
misery. In 1869, with a view of re
flagrant evils that had thus arisen,
of them was taken from the Excise
and placed in the hands of the magi:
however, have no power to sunn
existing before 1869, except on the
the holder's character, or that the
orderly and the resort of thieves,
applicant has previously forfeited
that there is a disqualification i
by law. The same rules apply t
Although the magistrates have not g
new beerhouse licenses, and a ve
fallen in, there are still, as alread
1,323 beerhouses in the City of Ma

The square mark on the map
license for " beer to be consumed
mises." There are 592 such licer
and they are chiefly held by small
was not until 1882 that the magi:
any control over this branch of the

The licenses for the sale of " Be
to be consumed on the premises" ar
the symbol L. There are 283 in
Manchester, and the number has
late.

The licenses to sell " Sweets"—
so-called " British wines"—are
There are 27 of them.

Then there remains a miscellaneo
other licenses," which are mark
include those granted to wholesale c
Of these there are seventy-six.

This brief explanation will enable
ascertain the exact extent of the lic
tations to intemperance that exist in
taken as a whole, and in each part
sidered separately. The year 1886 w
extension of the city boundaries by
poration of Rusholme, Bradford, and
The total number of licenses then
and at the licensing sessions of 1888 t
The population of the City of Ma
estimated at 376,164, so that there i
seller to each 107 persons ; "persons
including old and young, men, w
children. There is a drink license
for about each group of twenty-fi
The revenue of the Manchester d
must be reckoned not by the hundr
the million. A glance at the map
these drinkshops are blotched all ov
of the city is sufficient to prove tha
trade must yield enormous profits.

But is it profitable for the communit

Nearly one half of the prisoners
the City of Manchester are drink
they fall into the hands of the police,
fall into crime through intemperar
than one half of Manchester pauperi
proved to be directly due to drink
A reference to the map will show th
the poorer districts where the drink
abound. They live, in fact, on the
should be expended upon the comfor
and in this way they damage every
contributes to the comfort and prosp
people. Publicans flourish where
children starve and pine. Turned
channels the money now wasted upon
and resulting in poverty, disease,
would give fresh prosperity to every
contribute to the public good.

What is the remedy ? The evil
black, and monstrous. How is it to be
By giving the people the power
themselves *against* the liquor traffic.
the licenses now in force in Manchest
granted in obedience to the popular
some of the drinkshops have been ope
flagrant and wilful disregard of the
the people. Why should anyone be
force a drinkshop upon an unwillin
tenancy ? Let the people in each dist
by means of a direct vote whether tl
these snares for the unwary to be pl:
way of sons and daughters, of hus
wives. This is the plan of the United
Alliance, which asks that the question
or no license shall be decided by the p
by the people alone.

When the power of the direct vote
the struggle between the people and
will be decisive. The foul blotches
that now disfigure the map will be sur
increased health, order, and comfort
have worn opposed to the interests of
are the trade interests of the public
removal of the most active agent of
disorder, disease, and crime, woul
boundless prosperity to the City of M
Why deny to the citizens the powe
the community from the evil of t
traffic ?

THE DRINK MAP OF MANCHESTER

The Drink Map of Manchester is a reproduction of an instructive document, prepared for the information and use of the City Magistrates. It appeared in the *Manchester Guardian* on January 5th, 1889, and is now reprinted by permission. It shows at a glance the exact position of the liquor traffic in the great capital of the textile industries. There are some fifteen different kinds of licenses for the sale of intoxicants recognised by the law; but for all practical purposes the various branches of the liquor traffic may be accurately indicated by the use of the six distinctive signs employed in the accompanying map.

First, we may take the licensed victuallers. There are 484 hotels and public-houses. These are indicated on the map by triangular symbols. The Justices have absolute power to refuse any application for a new license. They have also absolute judicial discretion to refuse the renewal of a license previously granted. This may be done not only on the ground of the misconduct of the license holder, but for any other reason satisfactory to the magistrates, so long as they consider each case judicially upon its merits.

There is a power of appeal to the Court of Quarter Sessions, which in this particular district has gained an unenviable notoriety by the manner in which it has disregarded the earnest wishes of neighbourhoods anxious to be relieved from the curse which the liquor trade brings into every district where it gains entrance. The applicant for a license may appeal on a point of law to the Court of Queen's Bench. The powers of the Justices in regard to transfer are in substance the same as in regard to new grants and grants by way of renewal.

Beerhouses are indicated on the map by a circular dot. There are 1,323 such houses. The Beerhouse Act of 1832 [*sic*] is a monument of the folly of legislators. It was passed, in defiance of warning and opposition, with the avowed object of promoting sobriety! Give people plenty of opportunity to drink beer, it was said, and they will cease to drink spirits. The opportunity was given. On the mere payment of a fee to the Excise any one could open a beerhouse, and they quickly swarmed in all great towns, and caused an immense increase of drunkenness and misery. In 1869, with a view of mitigating the flagrant evils that had thus arisen, the licensing of them was taken from the Excise authorities and placed in the hands of the magistrates, who, however, have no power to annul a license existing before 1869, except on the grounds of the holder's character, or that the house is disorderly and the resort of thieves, or that the applicant has previously forfeited a license, or that there is a disqualification of the house by law. The same rules apply to transfers. Although the magistrates have not granted many new beerhouse licenses, and a number have fallen in, there are still, as already indicated, 1,323 beerhouses in the City of Manchester.

The square mark on the map indicates a license for 'beer to be consumed off the premises.' There are 392 such licenses in force, and they are chiefly held by small grocers. It was not until 1882 that the magistrates had any control over this branch of the liquor trade [a nod to the *Over Darwen* decision].

The licenses for the sale of 'Beer and wine to be consumed on premises' are marked by the symbol L. There are 283 in the City of Manchester, and the number has increased of late.

The licenses to sell 'Sweets' – that is, the so-called 'British wines' – are marked O. There are 27 of them. Then there remains a miscellaneous group of 'other licenses,' which are marked X, and include those granted to wholesale dealers, &c. Of these there are seventy-six.

This brief explanation will enable anyone to ascertain the exact extent of the licensed temptations to intemperance that exist in Manchester taken as a whole, and in each part of it considered separately. The year 1886 witnessed the extension of the city boundaries by the incorporation of Rusholme, Bradford, and Harpurhey. The total number of licenses then was 2,606, and at the licensing sessions of 1888 it was 2,585. The population of the City of Manchester is estimated at 378,164, so that there is one drink-seller to each 107 persons; 'persons,' of course, including old and young, men, women, and children. There is a drink license in existence for about each group of twenty-five families. The revenue of the Manchester drink traffic must be reckoned not by the hundred, but by the million. A glance at the map showing that these drinkshops are blotched all over the face of the city is sufficient to prove that the liquor trade must yield enormous profits.

But is it profitable for the community?

Nearly one half of the prisoners arrested in the City of Manchester are drunk even when they fall into the hands of the police. How many fall into crime through intemperance? More than one half of Manchester pauperism has been proved to be directly due to drinking habits. A reference to the map will show that it is in the poorer districts where the drinkshops most abound. They live, in fact, on the money that should be expended upon the comforts of home, and in this way they damage every trade that contributes to the comfort and prosperity of the people. Publicans flourish where wives and children starve and pine. Turned into useful channels the money now wasted upon intoxicants and resulting in poverty, disease, and crime, would give fresh prosperity to every trade that contributes to the public good.

What is the remedy? The evil is palpable, black, and monstrous. How is it to be removed? By giving the people the power to protect themselves *against the liquor traffic*. Not one of the licenses now in force in Manchester has been granted in obedience to the popular will, whilst some of the drinkshops have been established in flagrant and wilful disregard of the wishes of the people. Why should anyone be allowed to force a drinkshop upon an unwilling constituency? Let the people in each district decide by means of a direct veto whether they desire these snares for the unwary to be placed in the way of sons and daughters, of husbands and wives. This is the plan of the United Kingdom Alliance, which asks that the question of license or no license shall be decided by the people and by the people alone.

When the power of the direct veto is gained the struggle between the people and the publican will be decisive. The foul blotches of drink that now disfigure the map will be removed, and increased health, order, and comfort will show how much opposed to the interests of the public and the trade interests of the publican. The removal of the most active agent of poverty, disorder, disease, and crime, would give a boundless prosperity to the City of Manchester. Why deny to the citizens the power to protect the community from the evil of the drink traffic?

This third map has proven to be the hardiest. Beautiful examples of the original 4,000 printed copies survive in the collections of Manchester University's and Chetham's libraries, as well as on the wall – framed and visible through the window – of the Crown & Kettle pub in central Manchester.

The main point, introduced at the very end – and therefore perhaps not as effective as it could have been – was to argue the so-called 'local option' or the 'people's veto'. This widely bandied-about idea was a misunderstood concept even among ardent and seasoned activists. Sir Wilfrid Lawson himself at first argued one extreme: that it meant the power of the local population to *allow* a licence; meaning the presumption should be that no licences would be granted unless a majority of locals demanded it. If Lawson had his way, public houses would only be allowed if a majority of people who lived near them insisted upon them, not merely not voted against them. This was a ridiculously high bar, as anyone who followed Lawson would expect him to propose. The more commonly understood meaning of local option was that a licence would be granted unless enough locals turned out to vote against it. Some argued it meant a specific percentage of people, usually 25 or 30 per cent. There was also a go-dry version where a two-thirds majority was needed to bind a whole area. On top of disagreeing about the meaning of 'local option' and how to implement it, there were squabbles about its priority in competing for temperance resources of money, time and attention, with the primary competition being the Sunday closing legislation favoured by many temperance activists who resolutely asserted it should be the first regulation pursued, with local option considered only once Sunday closing laws were secured. In 1880, a Local Option Resolution carried for the first time in the House of Commons, although it would not make it all the way through. The infighting led to neither passing.

The rationale for having stricter rules around liquor licences was made on behalf of the public interest. In the same way, Parliament

28 Illustration of a drunken man outside a gin shop in Manchester. From *The Band of Hope Review*, 1870.

passed licensing laws, but it was the local magistrates who granted or refused individual licences. The local option movement would have allowed for the proximity of those living nearby to be a factor, and possibly the deciding factor, upon which magistrates could base their decisions. The most open interpretation of a local option meant that anyone in a neighbourhood (including those who usually didn't have a voice and could not vote in elections, like women and non-property-owning locals) could influence whether more places to imbibe intoxicants were needed in their near surroundings. Of course the 'drink trade' did not like the concept of a people's veto at all and fought it at every turn.

THE 'MANCHESTER EPIDEMIC'

Rivalling Liverpool and London, Manchester was also an infamously hard-drinking city. It already had a high rate of what was known as 'alcoholic neuritis', which is essentially a type of poisoning from too much imbibing and presents itself with rashes and shooting pain in the hands and feet. Suddenly in 1900 the cases skyrocketed. At first it was blamed on a sharp increase in drinking, but cases soared even among those who claimed moderation, and people were dying. A locally born physician, Ernest Reynolds, made the connection to beer and he examined samples from several breweries and public houses where his patients had been drinking. He found that the beer had dangerous

29 Engraving of the Gin Palace from *The Drunkard's Children* by George Cruikshank, 1848.

levels of arsenic. It would not have been added intentionally for flavour, presentation or any other reason, so further investigation ensued. It turned out that the arsenic had been passed on by the local breweries' sugar supplier Bostock & Co. They had dispensed to hundreds of breweries that were trying to save money by supplementing low-quality malt with sugar, a cheap alternative source of nutrients for the yeast to eat which was now legal but still controversial. The sugar supplement was most often used in the cheapest beers, so the poisoning hit the working classes the hardest. It turned out that a step in the sugar refining process was tainted with arsenic. Symptoms included discolouration around the neck, abdomen, armpits, feet and private parts. It didn't

30 Engraving from *The Drunkard's Children* by George Cruikshank, 1848.

help that doctors simply believed their patients were dirty and all they needed was a good wash. Within months reports of the poisoning spread across northern and midland England, although unfortunately due to its outbreak origination point everyone called it the 'Manchester epidemic'. Further investigation revealed that there were other sources of arsenic in the area, some unwittingly coming from the breweries and maltsters themselves, and had likely gone on for years. The whole dilemma brought into question the very existence of 'alcoholic neuritis' and challenged the approaches of clinical physicians' assumptions about their patients – that they were dirty, poor alcoholics – which prevented them from properly diagnosing the real and much deadlier problem earlier. Many temperance advocates were quick to spotlight only the alcohol part of the problem though, because it suited their agendas.

31 The most colourful of the known drink maps. Details from *Map of Manchester indicating the Licensed Victuallers, Beer, Wine and other Licensed Houses*, 1889.

FROM RED TO RAINBOW

In contrast to the red-on-red Drink Map of Manchester published by the United Kingdom Alliance, a much more colourful *Map of Manchester indicating the Licensed Victuallers, Beer, Wine and other Licensed Houses* was published that same year by Henry Blacklock & Co. of the better-known *Bradshaw's* railway guides (FIG. 31). Railway stations had their own rules and were outside of the general licensing scheme. Essentially, passengers could always drink at a railway pub, although drinking on the train itself was outlawed in 1881. While Henry Blacklock himself had died in 1871, at the fairly young age of fifty-two, his son Charles inherited a partnership share and took the business over entirely in 1885. The dots on this map pop from the surface with seven different colours – blue for licensed victuallers, red for beer consumed on the premises, orange for beer consumed off-premises, light green for beer and wine consumed on the premises, black for sweets on-premises, white (which does not stand out much against the white background of the map) for other licences, and a rust-coloured circle with an *X* through it for breweries. The bright markings give the viewer a faster and richer view of the licensing situation than its red-on-red cousin. It is a wonder that more anti-drink maps did not use additional colours given the intense impression of this image. Perhaps displaying so many colours was too expensive for those who didn't own a printing business, or maybe it's just too pretty and temperance societies preferred to continue the metaphor of likening a rash of red to sickness. As with many drink maps, the pattern resulting from so many scarlet spots was often compared to disease, to paint the evils of drinking as synonymous with dire afflictions – as Norwich's drink map did when it referred to them as 'plague spots', as you will soon read. The media were quick to embrace the sickness analogy. A few days after the all-red *Drink Map of Manchester* first appeared in the *Manchester Guardian*, the 9 January 1889 *Burnley Gazette* in Lancashire noted 'as it is represented in the drink map of Manchester, and as it exists in every town and village throughout the country: a traffic which like a foul and loathsome disease is spreading misery and ruin among the homes of the poor, and leaving the poisonous trail of its horrible infection in the lives of unborn generations.'

THE DRINK MAPS OF NORWICH

Unlike the drink maps of Manchester, which were all made within a year and delivered first to magistrates and then shared with the broader public, the drink maps of Norwich were published over a considerable period of time. The twenty-five-year span of their appearance, along with records of the numbers of licences granted and denied over that same time in the town, provides a chronicle of their effectiveness.

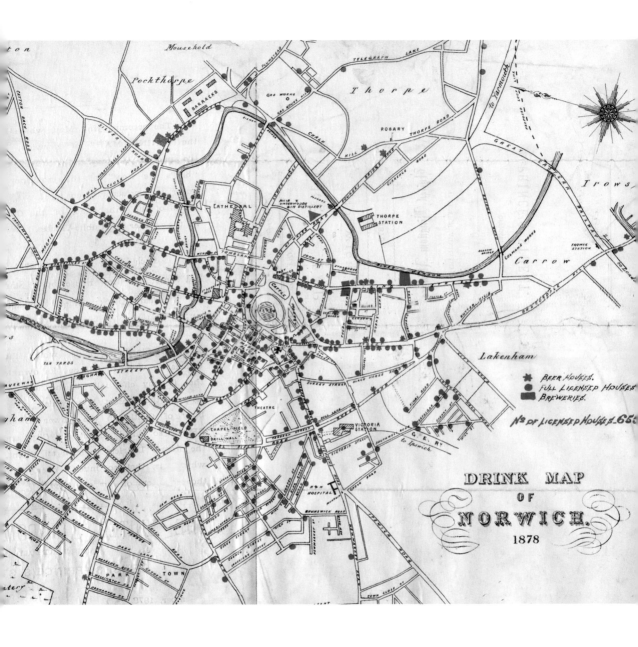

32 This is the first known drink map to use the words 'Drink Map' in the title and on its face. Note that the *Drink Map of Southampton* was published later that same year. *Drink Map of Norwich*, 1878.

Norwich's first and last drink maps act as bookends to the quarter century of the most prolific drink-map production, with a middle version acting as a litmus test of their success. It was the 1878 *Drink Map of Norwich* that was likely the very first of any temperance map to use the words 'Drink Map' in the title and on the face of the map itself (FIG. 32). Earlier maps, even those referred to as drink maps in newspaper articles about them, usually had the words 'licenses' (or 'licences'), 'public houses' or 'spirit dealers' in the title but not the words 'drink map', such as the giant 1840 *Map of Old Licences at Newton Heath* and the 1872 *Map of the City of Aberdeen shewing the numbers & position of the Licensed Grocers & Spirit Dealers* (see Chapter 1). At the time that the 1878 *Drink Map of Norwich* was circulated by the local auxiliary of the United Kingdom Alliance, there were roughly 78,500 people living in Norwich. According to the face of the map, which has a legend assigning circular, square and star-shaped symbols for different types of licence, there were 655 drinking establishments in the town at the time. In other words, 120 thirsty people per licensed house. The legend does not provide a tally of each type of licence – it leaves the reader to count them. And the total number does not include wine merchants or grocers, so there were certainly more marks that could have been added. Like other anti-drinking maps, it has text in the form of letter-press directly on it, in this case on the back, explaining the reason why the map had been created. It reads:

> TO THE CITIZENS OF NORWICH!
> The accompanying Map, which may be fitly designated the 'DRINK MAP' of Norwich, has been published for the purpose of presenting to the Citizens of Norwich, a 'bird's eye view' of the provision which has been made under the present Licensing laws, to meet the presumed wants of the Inhabitants of the City and Visitors thereto, in the matter of Public-Houses.
>
> Each Full Licensed House is represented by a red Spot; a Beer House by a Star; and all Breweries by a square Block;

Wine Merchants or Grocers who sell Wines and Spirits are not represented.

The Startling Picture which thus meets the eye, is sufficiently vivid to tell out its own lesson, and it is hoped that it may effectually teach those who look upon it, the folly of the present law, which gives a non-representative body of Gentlemen the power to place these PLAGUE SPOTS so thickly about our City, and especially to crowd the Poorer Districts to such an extent, as to show at once why it is, that so much Poverty and Misery abound in those quarters of our old City.

Its brief message speaks directly to the citizens about insufficient licensing laws, likens the spots on the map to ill health and blames alcohol for poverty and misery – specifically noting the clumping of dots in the poorest areas. The exclusion of shops and grocers that sold wines and spirits may have been to better highlight the impact of the Beerhouse Act.

This first Norwich drink map was printed in a run of 4,000 copies and sent to every Member of Parliament, to magistrates, and to men who were thought to matter. It was folded with 'Drink Map of Norwich' facing up as its title page, prominently displaying the national and local temperance groups responsible for publishing it. While the Norwich map touched on two topics that almost all drink maps raised, the social and legal aspects of the current alcohol access situation, later maps of other towns gave much more space to legal developments and described what the viewer could do to take action and achieve change. Here, on this pre-*Over Darwen* decision letterpress, the message is more of an awareness raiser. The goal seems to be to blame alcohol for crime and poor neighbourhoods, but there is no advice about what the angry and shocked viewer could do to alleviate the problem that had been identified. Explicit guidance for further action became a key element on later maps. Discussion of the map's publication in contemporaneous newspaper articles referred to the momentum growing in Parliament in support of at least some restrictions on alcohol sales – noting that

'Sunday laws' (prohibiting or limiting hours of sales on Sundays) had passed in Wales and Scotland, and that therefore England could hope to be next. In addition to the Norwich drink map being sent to decision-makers, there had been over seventy public meetings in the county alone, 2,500 leaflets circulated about the UKA, hundreds of sermons delivered, and the opening of public-house distractions like the Victoria and Alexandra Cafes, which proved to be competition for public houses. Newspaper accounts noted that Sir Wilfrid Lawson's latest temperance legislation was garnering more support in Parliament, thanks in part to help from Norwich's local Member of Parliament.

Norwich's beloved Member of Parliament was Jeremiah James Colman, whose family dynasty is universally remembered for mustard, and he himself was official mustard maker to Queen Victoria (although later he served in numerous influential roles including mayor, sheriff and magistrate). Colman mentioned the map in a meeting about the opening of a new coffee house. He told the crowd that they should not worry about the coffee house being filled with teetotallers (who were seen by some as extremist) because the contemporary temperance movement emphasized moderation instead of complete abstinence. He expressed hope that one day a map would be published showing the coffee houses of Norwich rather than the public houses marked on the recent United Kingdom Alliance map (referring to the 1878 *Drink Map of Norwich*). His assurance suggests that not everyone was on board with severe temperance views. Yet, by making room for the extremist teetotallers, the UKA created a space where moderation and some restrictions could be seen as a reasonable middle ground. Colman supported Lawson's Permissive Bill and the concept of 'local option', which is alluded to on the map where the rash of red spots is blamed on the 'non-representative body of Gentlemen'. Also known as magistrates.

Fourteen years later, a second *Drink Map of Norwich* was published by a different organization, the Gospel Temperance Union, founded in

1880. The GTU had overlapping leadership with the UKA, which apparently decided not to publish the map again. This second map used the same base map, type style, symbols, colours and visual approach as the first one. The differences are the reduced number of licensed houses and the lack of written message on the back. The reduction of spots was small: from 655 in 1878 to 631 in 1892 – just 24 fewer; less than 4 per cent. And under close examination, the number of blotches had indeed gone down on the edges of town, yet had become more concentrated in the centre of town, so visually it maintained its alarming appearance. Other maps in the County of Norfolk, like that of the Borough of King's Lynn of the same year published by the King's Lynn Vigilance Committee, used yardage circles radiating out from a point in the centre of town so readers could easily count how many triangles, squares and spots were within each circle and therefore how concentrated the public houses were. Newspaper articles at the time celebrated the map's printing.

Great Yarmouth, another Norfolk town, printed a drink map in 1903 at the request of the local licensing justices. Like other drink maps, the beer shops were marked, and on this one the measurement included how many fully licensed houses and beer shops were within a five-minute walk. There were 110! The shop fronts must have been narrow and on both sides of the street. This map is not known to have survived; however, an earlier map titled the *Map of Great Yarmouth Showing the Position of Licensed Houses* was published in 1882 and resides in the Norfolk Record Office.

The final *Drink Map of Norwich* was circulated in 1903, eleven years after the second one. It was published by three organizations jointly: the Joint Brewster Sessions Committee of the Church of England Temperance Society, the Norfolk & Norwich Gospel Temperance Union and the Temperance Divisional Council. It showed 615 houses – a trimming of just sixteen since the last map had been printed and a reduction of fewer than two alcohol destinations per year since the first version had been circulated twenty-four years earlier. This *de minimis* decline in the number of public houses could be used to argue that drink maps as tools to curtail drinking simply did not work. Just as the first *Drink Map of Norwich* was also the first to have 'Drink Map' in its title, the third and last anti-drinking map of Norwich was the last of such maps to have 'Drink Map' on its face, providing a parenthetical closing to the era of thought-provoking, and thirst-provoking, maps.

A BEERY TOWN

Norwich sits at the centre of desirable barley crops and is positioned atop chalk, which provides high-quality brewing water. The conditions are ideal for making beer, and from the late 1700s to 1836 breweries grew from 9 to 27, in addition to public houses that brewed on premises. In 1845 there were over 500 public houses in the city, inspiring its notorious slogan, 'a pub for every day of the year', although there were actually many more. Perhaps a pub for every afternoon and evening would have been closer to reality. Not everyone thought this was a positive image. One of Norwich's most vocal and ardent abstainers was George White, who served as secretary of the Norwich auxiliary branch of the United Kingdom Alliance before becoming president of the newly formed Norfolk & Norwich Gospel Temperance and Blue Ribbon Union, founded in 1901, which co-published the third *Drink Map of Norwich*. White was big on numbers. When he eventually became the Member of Parliament for North West Norfolk in 1900, he launched the 'Million New Pledges Crusade' in a nationwide effort to increase the number of complete abstainers from alcohol. Although he did not reach a million, the scramble created a frenzy. His extreme views relied on morality instead of law, which may explain why he left the UKA for an even more rigid organization.

TIED NORFOLK

The 'tied house' system led to takeovers
and consolidation, eviscerating small
breweries and leaving consumers with
few choices. A tied house is the opposite
to a free house, which can sell beer
from wherever it wants. The tie was
essentially a finance mechanism which
provided up-front liquidity to publicans
so they could get their businesses going
or make improvements, and a reliable
market for brewers so they didn't have
to compete with other beer makers for
space at the bar. Or at least that was
supposed to be the idea, but as time
went on a few large breweries – six to
be exact – owned or controlled 90 per
cent of the tied houses in the UK. The
interpretation of these developments
depends on who profited. For example,
an exhibit at the recently closed
National Brewing Centre in Burton
upon Trent referred to the Victorian
growth of the tied system as large breweries 'rescuing' public houses. That is
one perspective. Long after drink maps
were no longer being made, in 1989, a
review of the tied system determined
that it was, in fact, a monopoly and
mandated by way of the 'Beer Orders',
a cap of 2,000 tied pubs per brewery.

33 The shape of the medal is
called the 'Crookshank Cross' in
honour of the former president
of the Soldiers' Total Abstinence
Association. India Temperance
Association six-months-sober blue
ribbon, 1893.

This was at a time when just one of the Big Six breweries, as they were known, owned over 7,000 pubs. In some towns there were many pubs but only one choice of beer. In Norwich, in particular, by 1875 there were only seven breweries left thanks to the tied-house scheme, although the number of public houses continued to grow – all tied to a few national breweries. Eventually, larger enterprises gobbled up nearly all of the remaining brewing locations until Norwich was left with just one, Watneys; and then even they closed in the 1980s. Norwich's beer scene has since revived.

Norwich has preserved its brewing history within the Museum of Norwich at the Bridewell. There is a permanent exhibit in this museum of the town's beery past, including a transplanted bar, drinking vessels, photos of historic pubs and brewery paraphernalia. It even has a plastic-coated oversized copy of the 1892 *Norwich Drink Map* exhibited with a hand-held magnifying glass so visitors can get their bearings.

SHOCKING TEMPERANCE TOOL
OR BEAUTIFUL PUB GUIDE?

Visually, all of the Norwich drink maps are rather decorative compared to other towns' maps, which lack such elaborate compass roses, elegant cartouches and fancy scroll work at the edges unless such flourishes were part of their original base maps. While the map refers to itself as a 'bird's eye view', it is actually not one, as that would be the view from a distant height looking down at the area with some perspective and angle – as a bird flying over would see the ground, or a human standing on a distant hill. All three drink maps of Norwich are a view from directly overhead, like most paper maps.

The drink maps of Norwich truly show the arc of influence of maps meant to discourage drinking in this era. There is little doubt that at first, especially when shown to magistrates to present information otherwise seen only in the form of dreary lists, there was an 'ah ha' moment upon viewing them. But as time went on and the maps were also used to try to influence the public, a level of scepticism arose. The accuracy of the maps was questioned when people could not identify their favourite watering holes, even though the maps themselves often contained a warning that the marked locations were estimates based on police records. The police themselves were doubted because (notoriously in Liverpool but elsewhere as well) it was assumed that police were in the pockets of pub owners. Each successive Norwich map, with the same intent to shock as the 1878 version but showing little difference in the number of imbibing options, actually undermined the argument and revealed that the strategy of using drink maps to organize citizens to reduce the number of public houses was not as effective as originally hoped.

CENTURY'S CLOSE

By the end of the century, after battles on several fronts for social reforms addressing the public welfare of the working classes and the poorest people, progress had been made. Literacy had reached 90 per cent for both men and women, more people – although still only men – could vote, and recreational activities were firmly established, such as cycling, swimming, fishing and birdwatching. Overall health had improved and alcohol consumption had come down. A man called Charles Booth made it his life's work to measure these changes in successive sets of multiple-volume treatises compiling reflections on the state of society. The last map in the back pocket of the last volume of his last series was not one of the poverty maps he is most famous for, but instead a drink map of London.

RECOMMENDED BY
LADY HENRY SOMERSET,
WEST END PHYSICIANS,
AND MANY OTHERS.

NO INJECTION.

PRIVATE WAITING
ROOMS.
NO PATIENT EVER MEETS
ANOTHER.

THE SECRETARY, OPPENHEIMER INSTITUTE,
231 & 232, Strand, London.

President:
THE LADY HENRY SOMERSET

Directory
OF
INEBRIATE HOMES
FOR
WOMEN.

MINISTERING

"Not to be ministered unto,
but to minister."

BY

LADY HENRY SOMERSET.

❋

PUBLISHED (FOR FREE)
National British Women's Temperance Association.
BY THE WHITE RIBBON COMPANY, LIMITED,
12, MEMORIAL HALL, FARRINGDON STREET, LONDON, E.C.

To Comfort and Help the Weak-hearted; and to
Raise up them that Fall.

**The Industrial Farm Colony
for Women, Duxhurst.**

A
WIDER OUTLOOK.

BY

LADY HENRY SOMERSET.

HOW TO ORGANIZE

TEMPERANCE MEETIN

WITH A PREFACE BY

LADY HENRY SOMERSET
President,
British Women's Temperance Association.

CHAPTER SIX

THE END OF THE
DRINK MAP ERA

Two prominent nineteenth-century advocates of social and alcohol reform relied on maps to make their points, and, while these contemporaries may never have met, they had a lot in common. They both came from wealthy families, and both used their positions and money to advocate on behalf of the poor. Before describing their cartographic contributions, it will be helpful to know how they came to offer their maps. One was a man and therefore accustomed to a receptive, respectful audience ready to believe him. The other was a woman, who fell from grace in a social scandal before she gained her independence and launched a life as a dedicated social reformer.

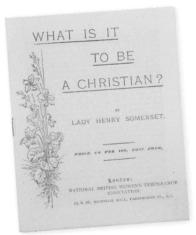

34 A selection of pamphlets on temperance by Lady Henry Somerset (c. 1895). Lady Somerset used her leadership position in the British Women's Temperance Association to promote live-in alcohol treatment centres for women, including her own at Duxhurst. She shared what she learned from her vast number of speaking experiences around the world and lessons from other successful groups to teach British women how to coordinate local temperance efforts effectively.

ISABEL SOMERSET

Lady Somerset (née Isabel Somers Cocks) was born in 1851, the same year that Neal Dow's Maine law prohibiting the sale of alcohol was first passed and a couple of years before the United Kingdom Alliance was founded. She married Henry Somerset in 1872. It was a celebrity wedding. Alfred, Lord Tennyson sent the flowers she carried. The prime minister, Benjamin Disraeli, attended along with hundreds of others. Money had married rank for an ideal union.

Or so it seemed. Isabel and Henry had one child before, it is believed, she caught him in bed with a teenage boy and did something to make her a social outcast for the rest of her life. She sued him. Not for divorce, because she was against that on religious grounds. But for custody of their four-year-old son. By law, custody was presumptively with the father, so she was forced to disclose a reason that the law would defer to. Meaning she had to out him. Ever since the implementation of Henry VIII's 1533 Buggery Act, men having sex with men had been punishable by death in England, only being reduced to life in prison a few years earlier in 1861. Lady Somerset prevailed in the custody matter (decided by the same Justice Field who would, a few years later, help decide the ruling in the *Over Darwen* case; see Chapter 3) and Henry left for Italy. Her fall in society for not looking the other way was catastrophic. She realized just how low she had plummeted when, while visiting her sister with whom she was quite close, a caller arrived. Her sister begged her to leave immediately – by the back stairs.

Eventually, however, Lady Somerset rose up the ranks of the temperance movement to become the president of the British Women's Temperance Association in 1890, having signed a total abstinence pledge in 1885 following the suicide of an intoxicated friend. She had a gift for oration that commanded audiences all over England, Scotland and Wales and in major American cities. Her first speech in the United States in 1891 was before 2,000 people in Boston at the first

World Women's Christian Temperance Union's convention, a group she would eventually lead. She went on to speak to crowds all over the United States and met President Benjamin Harrison. She was a household name at the time. In 1895 she founded the Duxhurst Home for Inebriates, which aimed to treat habitual drinkers instead of punish them – still a new concept at the time.

THE CROWN GATHERS EVIDENCE: THE ROYAL COMMISSION ON LICENSING LAWS

Arriving rather late to the sobriety crisis, the Crown finally decided to take an official look at the drink problem in 1896 by establishing the first Royal Commission on Licensing Laws. It convened 123 meetings, beginning in May of that year and continuing through to July of 1898. Two hundred and fifty witnesses testified, and a formal report was published in nine volumes in 1899.

CIDER SENSE

Lady Somerset lived at Eastnor Castle, a large family estate in Herefordshire which had many cider-producing farms. At a time when other agricultural producers used 'cider trucks' (carts that provided cider to workers in the orchards, and then deducted it from their pay) she discouraged cider production. Because she regularly visited her tenants and listened to people, she understood the impact of her decisions and took steps to make sure livelihoods did not suffer. For example, to offset the lost profits usually made by local cider manufacturing she sent her gardener (an award-winning horticulturalist) to her tenants to share his expertise and also to distribute new plants, and arranged for a supply-chain mechanism to help people sell other kinds of fruit and eating apples instead of cider apples. This hands-on empathetic approach became her signature. The Truck Act of 1887 finally prohibited the paying of wages in cider after studies found rampant abuse of workers.

The Commission's report, sometimes referred to as the Peel Report on account of the name of its chairman Lord Peel, was heavily skewed towards the drink trade's view of licensing. The witnesses were mostly brewers, publicans and leaders of drink-trade associations, with a handful of temperance-minded citizens sharing their views. Major temperance figures and parliamentary liberals, including Sir Wilfrid Lawson, openly dismissed the biased evidence-gathering. His unwilling-ness to participate and his public mockery about how the content of the report was being compiled was widely followed. The report did include enough sober-leaning contributors to allow for a minority report to be written – led by Chairman Peel himself – that countered the majority report. This published divide diluted the resulting report to the point of making it meaningless. The conclusions were discredited among tem-perance advocates before it was even printed. In the evidence-gathering phase, Lady Henry Somerset was one of the few women to testify. And she brought seven carefully made drink maps for her interrogators to study, two of which will be discussed here (FIGS 35 & 36).

The seven maps Lady Somerset brought to support her statements were bound within the third volume of the final report marked as 'Evidence' and assembled in 'Appendix IV'; her transcribed testimony is in this same volume. The maps are of different neighbourhoods in London. The largest area covered is shown on 'Map 4' titled *Soho, St. Martins, St. Giles Etc* and the smallest area covered is a single street shown on 'Map 2' titled *Whitechapel Road, E.*, which stretches for exactly 1 mile. She pointed out the large number of public houses in poorer areas and the sharp contrast of there being nearly none in the wealthier areas. For example, she asks for two maps to be compared to prove that the wealthier parts of the city have the means and interest in keeping alcohol out of them, 'Map 5' titled *Belsize Park, South Hampstead* and 'Map 6' titled *St. Mary, Somers Town and St James, Hampstead Road, N.W.* She emphasizes that her primary reason for submitting them is

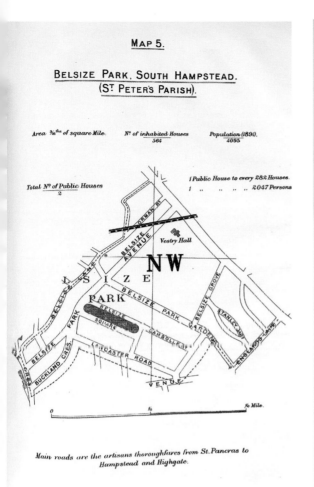

35 A map of the St Peter's Parish area of London (Belsize Park) showing two public houses. *Royal Commission on the Liquor Licensing Laws, Evidence, vol. III, Appendix IV, Map 5*, 1899.

36 A map of the area around Euston Station showing thirty-seven public houses. *Royal Commission on the Liquor Licensing Laws, Evidence, vol. III, Appendix IV, Map 6*, 1899.

to demonstrate that the wishes of the neighbourhoods are completely disregarded when it comes to licensing places to purchase alcohol, and that 'the only places which are comparatively free are those places which are rich enough to maintain an effective opposition to their establishment.'[1]

Lady Somerset was invited to testify because of her reputation for charitable works and a devotion to sobriety. By the time the Royal Commission convened, the home she founded was a respected and known success story. Highlights taken from Lady Somerset's testimony reveal an extensive understanding of the subject and a clever wit. For example, when asked about the drink trade supporting beauty contests among bar maids, she replied:

> I think the very fact that you have to seek out good looking and beautiful girls … shows that you have to produce as many attractions as possible in order to push the trade. I do not hear of beautiful female telegraphists being required for people to send more telegrams.[2]

Returning to the maps she submitted, her interrogators tried to insist that areas where people come into town just for the day to work could increase the needs of the neighbourhoods. She was ready for this, and pointed out that Maps 5 and 6 are of similarly sized neighbourhoods and she asked that they be viewed side by side to show that one is rich, one working class; one has just two public houses, while the other is scarred with thirty-seven of them. When the interrogator tried to say that working-class people did not walk through the wealthy neighbourhoods, she reminded him that they work for the rich in that area and there is a high road, yet there is nowhere to get a drink. She compared the main thoroughfares of each map, walked by people heading to the trains, but only the one in the working-class neighbourhood is crowded with places in which to consume alcohol. She provided further encouragement in her testimony to study the drink maps she brought, which included Map 1, *Soho District West*; Map 3, *Fitzroy Square District West*; and Map 7, *Pimlico, Belgravia Etc.*

She patiently answered the questions of fifteen examiners, mostly solicitors, judges and Members of Parliament. Her knowledge and experience relating to alcohol and novel ideas of treatment came from her travelling widely and collecting ideas from the United States, Europe and Scandinavian countries. It's clear from the tone of the queries and follow-up responses that Lady Somerset was respected by the men seeking her input and that her knowledge was valued. Unlike most witnesses, an entire day was devoted to her being interviewed on a range of topics, from New York barmaids and women-only public houses in Wales to the futility of petitions to influence policy – and on the use of drink maps to influence and educate magistrates. The result was twenty-one pages of evidence, which was much more than most witnesses, in addition to the seven maps she submitted.

When asked what rule she believed the licensing justices followed when making decisions, Lady Somerset testified:

> They would, I suppose, look at the map and see whether there was a public-house in close proximity? – I think as a rule a foot rule [a foot-long ruler] and an Ordnance map are the two instruments that are thought most necessary to see exactly how far a public-house is to be planted from another, rather than what are the absolute requirements of the place.[3]

She presumed that all magistrates had a drink map to consult when making their decisions, and she may have been right. She also clearly believed that distance between licensed houses alone was not the best way to determine if the needs of a neighbourhood were being met, and that instead the people should have a say.

CHARLES BOOTH

Charles Booth was born to a wealthy ship-owning family in 1840. He could have joined the family business, but after observing poverty first-hand in London his mind was committed elsewhere. In 1885 – the same year that Lady Somerset founded a treatment centre for women inebriates – Booth set about determining the number of people living

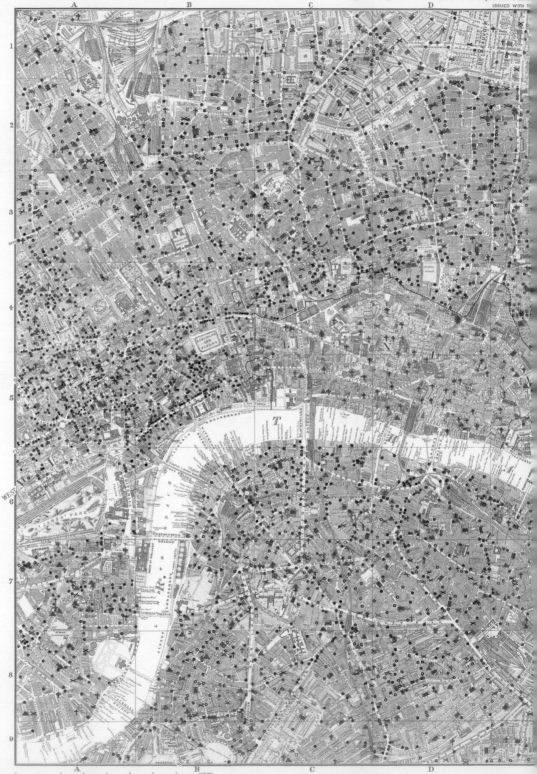

Houses Licensed for the Sale of Intoxicating Drinks.

OUR OF THE PEOPLE"

G   H   I   J

EXPLANATION OF SYMBOLS.

PLACES OF WORSHIP.

Church of England   Nonconformist Church
Church of England Mission   Nonconformist & Unsectarian Missions
Roman Catholic Church   Jewish Synagogue

PUBLIC ELEMENTARY SCHOOLS.

Board School   Voluntary School

HOUSES LICENSED TO SELL INTOXICATING DRINKS.

Fully licensed house
Beerhouse with "on" and "off" license
Beerhouse with "off" license
Grocers with license to retail wine, beer or spirit in bottles
Restaurant with wine, beer or spirit license, but without a "bar"

LICENSED HOUSES IN THE CITY OF LONDON ARE OMITTED.

Note as to Places of Worship:
A "Place of Worship," as here defined, is a building used primarily for religious purposes, and in which a public service for adults is regularly held on Sundays.

Note as to Public Houses:
Fully licensed houses are those licensed to sell beer, wine or spirits, to be consumed either "on" or "off" the premises.
Beerhouses with "on" or "off" license are those which sell beer, or beer and wine, to be consumed either "on" or "off" the premises.
Beerhouses with "off" license are those which sell beer to be consumed "off" the premises.
Grocers are those who, in addition to their ordinary trade, are licensed to retail beer, wine, or spirits in bottles, or are wine merchants selling grocery.
Restaurants which have a "bar" are treated either as fully-licensed houses or as beerhouses, according to the nature of their license. If without a "bar" they are enumerated separately.
N.B.—There are certain licensed houses which do not fall exactly under any of the above categories; such houses have been included under that heading to which, in the nature of their business, they most nearly conform.
Chemists with wine licenses are not included, nor are dealers in wine, beer, or spirits in bottles just to be consumed on premises, unless they sell for, &c., as well.

1

2

3

DOCK TILBURY

IMPERIAL GAS

4

EAST INDIA DOCKS

5

LONDON DOCKS

POOL

Lavender Pond

WEST INDIA DOCKS

BLACKWALL BASIN

6

COMMERCIAL

MILLWALL DOCKS

7

8

GREENWICH REACH

9

F   G   H   I

Stanford's Geographical Establishment, London.

in poverty in London. He ended up proving there were even more people in this situation than anyone had suspected, over 30 per cent of those living in the city. The maps that resulted from his research, and for which he is most famous, are known as 'poverty maps'. It wasn't until the later years of his career that Booth added a map showing public houses, which curiously he did not overlay with poverty information or colouring. In the last book of *Life and Labour of the People in London, Final Volume: Notes on Social Influences and Conclusion* Booth shares impressions on drinking and clubs, public houses and licensing, among other things. To find the map, readers are directed to a *Map of Inner and East London (In a Pocket) (Showing Places of Worship, Schools and Licensed Houses)* (FIG. 37). This map reveals a shift in his focus from pointing out where the poor were located to how he believed poverty could be eradicated: by educating children, increasing morality (by means of religion) and decreasing drunkenness. Thus the map shows, in blues, reds and blacks, schools, churches and public houses respectively. Visually, the colours Booth chose for his map seem almost to downplay the licensed houses. It is the red markings for religious institutions that pop off the page, not the black marks identifying where to find intoxicating beverages. Perhaps he wanted to maintain the consistency of the black he used on his poverty maps to associate the colour with negative connotations, but at first glance one cannot help but think that there are a lot of churches in London. The bright blue symbols for schools also stand out more than the black drink circles. However, and as Booth explains in his accompanying book, he is more concerned with showing how densely packed together the marks are. This is not a map to glance at; it is a map to study. Still,

37 (*previous spread*) This map was folded into a pocket in the back of the last volume of Charles Booth's 17-volume *Life and Labour of the People in London* (London, 1899–1900). *Map Showing Places of Religious Worship, Public Elementary Schools, and Houses Licensed for the Sale of Intoxicating Drinks,* 1902.

though, there is another missed opportunity in the City of London, which at first appears to not have any licensed houses at all. In the fine print of the legend it is noted that they are omitted, and in the book itself Booth explains that they cannot be shown because the area would be a blanket of black. But surely a massive black splotch in the centre of the map would have had more impact than an area just showing schools and churches. It would have encouraged the viewer to seek an explanation. The omission suggests that Booth's goal in making the map was not necessarily temperance. The area seems spared of demon drink, and does not inspire one to delve further. Compare this map to another map of London where drinking outlets were the primary focus and inspiring temperance its clear motivation. *The Modern Plague of London: Showing the Public Houses as Specified in the London Directory* (FIG. 38) was published by the National Temperance Publication Depot, and subsequent versions were published in 1886, 1887 and 1900. Its pink dots look like hives covering Bacon's *Map of London & Suburbs.*[5] There was tension between Manchester and London in vying for geographical dominance of the temperance movement. The United Kingdom Alliance and its leaders commanded most of Britain yet played second fiddle to the National Temperance League when it came to London. Despite its name, the National Temperance League was quite focused on London, while the UKA (whose archives are housed in London today) worked on spreading its influence as far as possible – to Wales, Scotland, Ireland (then wholly in the UK) and nearly every town in England.

38 (*overleaf*) The National Temperance League was responsible for compiling a drink map in 1884, not surprisingly of London, showing the public houses as specified in the London Directory. It was published by the National Temperance Publication Depot, and subsequent versions were released in 1886, 1887 and 1900. The final one appeared as part of the social-science section of the Paris Exhibition. Its pink dots look like hives covering Bacon's Map of London & Suburbs. *The Modern Plague of London: Showing the Public Houses as Specified in the London Directory*, 1886.

MAP OF
LONDON
Showing the Public Houses
IN THE METROPOLIS,
being carefully compiled from the
LONDON DIRECTORY.

Public houses shown thus

Lady Somerset and Charles Booth's social research of late Victorian life in England revealed trends in the licensing loophole of private clubs, children's drinking connected to the class of their parents, compensation scheme beneficiaries, workhouse abuses and women's drinking habits. Their contributions would lead to pension schemes, bans on children fetching alcohol, and other protections.

PRIVATE CLUBS

Lady Somerset lamented that one way to get around licensing laws was to declare a private club. Clubs did not require licences to serve alcohol. She claimed to know of many where the sole purpose of the formation of a club was to drink. It's true that only a few maps even mention clubs – those of Birmingham and Derby, for example. Drink maps were usually compiled by reports supplied by police, and they only tracked licensed establishments. This may explain why the student pubs and college breweries, such as those of New College, Brasenose College and The Queen's College that were part of Oxford University, are not on that city's drink map. As the century went on, drink-related misconduct connected to clubs was followed more closely. Even though the clubs weren't licensed, police could clamp down on disorderly conduct. In this way some clubs got their attention and were fined and even closed. But for the most part they remained a loophole.

CHILDREN FETCHING ALCOHOL

It was common in Victorian England for children to go to the pub to fetch beer for their parents. Both Booth and Lady Somerset commented on a typical scene in poor and working-class neighbourhoods of children in line with empty jugs waiting outside public houses at dinner time to get beer for their parents. Lady Somerset complained that publicans lured children with candy or oranges, and testified that 'It is a very common thing to see children with sweets in one hand and the jug of beer in the other.'[4] A publican testified to the Commission that a sweet taste did not go well with beer and admitted he gave them to children because it prevented them from sipping off the top of their cargo; the sweetness discouraged drinking. Charles Booth refused to engage in the debate, arguing that it was the early familiarity with the public house that tempted children to drink throughout their lives. A police officer Booth interviewed explained that things were changing and public houses were becoming more respectable at the end of the century. He painted a sharp character:

> The modern publican is of a totally different type of the man of twenty years ago, with his white hat and black band, and his bull dog: the decayed prize-fighter type. The publican now is usually well educated; respectable, and a keen man of business, who can keep his own accounts in proper order, and fully realizes that it is to his interest that the law should be strictly observed in his house.[6]

Lady Somerset regularly spoke to groups such as the Order of the Rechabites about the crisis of child drunkenness in London, noting that in one year there were over 500 drunk children under ten years old picked up by law enforcement, and 1,500 under the age of fourteen. The recent legislation setting a minimum age for couriers of beer was widely ignored in the capital city.

COMPENSATION

Compensation was the idea that publicans should be paid when their licences were not renewed to help them bridge the inevitable gap in their livelihoods. (It's easy to return to poor Mr Kay of Darwen with this solution, but grocers – whose businesses could continue without the sale of alcohol – were not eligible under most proposed compensation schemes.) As a concept it was fiercely opposed by teetotallers who argued that just as slave owners did not deserve compensation for losing their immorally gained money when they lost their free labour, neither should the publicans from their reprehensible actions. There was also the straightforward argument that the statute itself said that a licence is valid for one year. More moderate temperance folks disagreed among themselves. Overall, with a few urban exceptions, compensation was not embraced. The rationale that prevailed was that a licence had no value after a year, and therefore there was nothing of value to pay for. Public-house licence owners had notice written directly in the statute itself that their licences were temporary and it should not be put upon the public at large to pay for the drink trade's intentional disregard of the risk of a licence not being renewed.

Liverpool was one city that embraced compensation, at least for a little while, and perhaps because of the layer of informal 'law' that prevailed there. Like Leicester, Liverpool had drink maps that were updated by hand over time. The drink trade had a stronger position here than most areas, and they could push harder on this issue. Liverpool had large neighbourhood drink maps for different areas called 'divisions', labelled 'C' Division, 'B' Division, and so on. These maps were held in each division's police courts and updated as licences were granted, denied or transferred. In the 'C' Division these changes were noted by hand on the maps with blue markings, added over time, crossing out the red spots of licensed drink outlets. The words 'Licenses Compensated & extinct' are written in cursive on the lower left corner directly on the map in the same blue used to cross off the dots (FIG. 39). In this thickly beer-shopped part of town near the docks there are over 100 red marks crossed out in blue. The map is dated 1899, but the Licensing Act relating to compensation was passed in

39 Note the hand-drawn markings crossing off the licences as they became 'extinct' when closed in exchange for compensation in a Liverpool police district. *Plan Shewing Licensed Public Houses within Police Divisions*, 'C' Division, 1889.

1894. This map was used to track the licensed establishments as they were closed. Newspapers reported swaths of public houses being closed through compensation schemes in other cities as well. At least the drink maps could finally be useful in tracking the closure of places to find a drink.

CHANGING DRINKING BEHAVIOURS

Near the end of the nineteenth century, Booth published his observations that even though more people were drinking, they were drinking less. It was becoming normal for men and women – and not just prostitutes – to enjoy drinking together in public. For an attitude adjustment after a day at work, to celebrate life events large and small, to counter the inhibitions of the 1800s. Gradually, there was no longer shame in entering a public house, and women were seen to treat each other to a glass of wine or beer. Booth proposed that women brought sociable behaviour to the public house because they were not there to get drunk, and that in fact outward drunkenness had fallen out of favour in all classes. Yet he also contended that drunkenness in women was more common after marriage, and that 'it is not until they get older that women become regular soakers'.[7]

LICENSING AND THE DRINK INDUSTRY

Both Lady Somerset and Charles Booth believed that licensing in the hands of magistrates had failed. The magistrates were not willing to go up against big brewers or the powerful drinks trade. The problem was that even though magistrates had the power to deny licences, they had only refused 46 of the 67,000 public-house licences in England and Wales in the five years since the *Over Darwen* case had been decided. There was a self-defeating rift within the temperance groups between extreme teetotallers who demanded nothing less than complete prohibition and those who believed in moderation and reform. This divide was ultimately to blame for why the temperance movement could not prevail using the legislative process or a legal mechanism against the drinks trade. Booth specifically declared that there had been a failure of temperance propaganda, to which drink maps belonged. Booth was right about that failure – although perhaps he did not mean maps specifically, as his own studies were popular in part because of the maps used to illustrate his points.

Just as historians point to many reasons for the upsurge of drinking at the start of the century, so too are there many reasons for alcohol consumption to crash from 1900 to 1920. There had been gradual positive changes over the years such as improved living and working conditions, entertaining distractions such as cycling, gardening, museums, libraries, easier travel and even homing-pigeon societies. The need for escapism decreased when life became more fun and people were healthier. Some changes were not as welcome but also influenced the decline of drinking, such as a new sobering tax to pay on alcohol and a war on the horizon. In true British fashion, a steady stoicism prevailed, and prevented an extremist solution such as the full prohibition of alcohol that Americans suffered from 1920 to 1933.

DRINK MAP LEGACY

Few drink maps were published by temperance groups after 1900, partly because other forces had succeeded in reducing the number of public houses and drinking in general. Twenty years and a war later the city of Edinburgh decided to use a drink map to encourage voting against proposed licensing laws (FIG. 40). The red dots are massive, the mapped area minimal, and the writing so large and brief that the meaning is understood in the same glance as the image. From a design perspective at least, lessons were learned. But as a fundamental tool of persuasion, this drink map – like those before it – arguably inspires more thirst than outrage.

THE HEART OF EDINBURGH

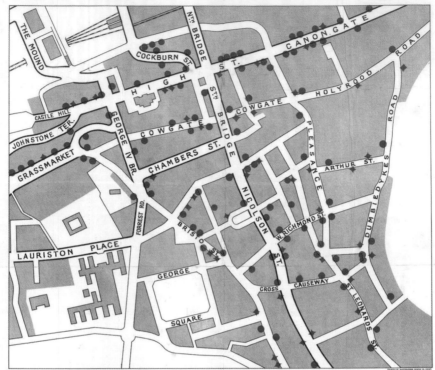

PUBLIC HOUSES.........●
LICENSED GROCERS...◆

ELECTORS!!
DO YOU WANT "NO CHANGE"?
VOTE LIMITATION OR NO LICENCE.

PUBLISHED BY EDINBURGH CITIZENS' "NO LICENCE" COUNCIL, 44 FREDERICK ST, EDINBURGH.

And now you know the story of drink maps. In the end it's hard to say if the maps had much of an impact on alcohol consumption, knowing that in some cities like Sheffield the number of public houses declined significantly after the circulation of its first drink map, while in others like Norwich the decline was minimal and in Leicester the number of drinking establishments actually rose. Drink maps were meant to be temporary. Some were so intentionally ephemeral that their publication dates prominently display not just the year, but also the month and sometimes even the day. In hindsight, it's rather incredible that any library or archive chose to add them to their collections at all. The maps may have failed in their original purpose, but as historical documents they have great value in what they reveal about drinking habits, the development of licensing laws and the tactics of the temperance movement in the UK.

40 This post-drink-map-era poster uses focused composition to drive attention to the intensely crammed crimson markings and words so few and large they can be read in an instant. *The Heart of Edinburgh*, 1923.

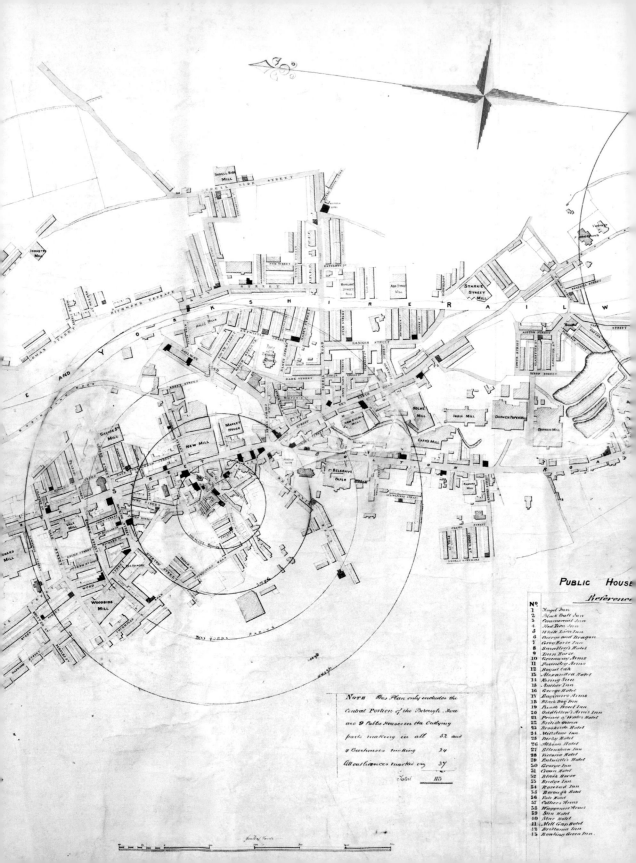

PUBLIC HOUSE

Reference

Nº	
1	Angel Inn
2	Black Bull Inn
3	Commercial Inn
4	Red Lion Inn
5	White Lion Inn
6	George and Dragon
7	Grey Horse Inn
8	Smalley's Hotel
9	Dun Horse
10	Grimshaw Arms
11	Fenniscowles Arms
12	Royal Oak
13	Alexandra Hotel
14	Rising Sun
15	Anchor Inn
16	George Hotel
17	Engineers Arms
18	Black Dog Inn
19	Black Horse Inn
20	Oddfellows Arms Inn
21	Prince of Wales Hotel
22	British Queen
23	Brookside Hotel
24	Millstone Inn
25	Derby Hotel
26	Albion Hotel
27	Ellenshaw Inn
28	Britania Hotel
29	Entwistle's Hotel
30	George Inn
31	Crown Hotel
32	Black Horse
33	Bridge Inn
34	Rosebud Inn
35	Borough Hotel
36	Fox Inn
37	Colliers Arms
38	Waggoners Arms
39	Sun Hotel
40	Star Hotel
41	Mill Gap Hotel
42	Brittania Inn
43	Bowling Green Inn

NOTE This Plan only includes the
Central Portion of the Borough there
are 9 Public Houses in the Outlying
parts making in all   52 and
4 Beerhouses making   24
All outlicences marked viz   37
                    Total   113

KEEP ON THE LOOKOUT

I found the drink map of Over Darwen by having a conversation with a stranger in a pub while drinking a local beer, which is arguably the quintessential pub experience.

Most of the maps mentioned in this book no longer exist. Their production and use have been confirmed primarily by reading about them in contemporaneous newspaper accounts, letters, records of legal proceedings and meeting minutes of temperance organizations. Many were printed in batches of 2,000–5,000, yet only around twenty-five are known to exist today. What happened to them? There are surely more out there, and those hidden treasures are what I am hoping the readers of this book will help to uncover. You just might find one hanging in the corner of your local pub or curled in a drawer of your attic. Please help contribute to preserving the knowledge and history of these maps by keeping a lookout. If you find one, share what you learn and add to this fascinating history by posting your discoveries on social media using @DrinkMapBook.

APPENDIX

DRINK MAPS
BY LOCATION & DATE

This is a list of drink maps encountered during the research of this book. Some were seen in person, some confirmed to be in archives, libraries or private collections, and some on public display. Others no longer exist, or perhaps remain to be found somewhere. They have been mentioned in letters, meeting minutes, legislative history, newspaper articles and other primary sources.

Aberdeen, 1872. *Map of the City of Aberdeen shewing the numbers & position of the Licensed Grocers & Spirit Dealers, within the Municipal Boundary, corrected up to July, 1872.*

Accrington, 1883. Published by the United Temperance Council of Accrington.

Belfast, 1883.

Bilston, 1892.

Birkenhead, 1899. Published by the Birkenhead Vigilance Committee.

Birmingham, 1876. Presented to the Town Council by MP Chamberlain in support of the Gothenburg system.

Birmingham, 1877. *Drink Map of Birmingham.*

Birmingham, 1885. Mentioned in testimony to the Royal Commission on Licensing Laws by Mr Barradale.

Birmingham, 1891. *Map of Licensed Houses, Birmingham.*

Brecon, 1893. *Drink Map of Brecon*, printed in *Brecon County Times*, 16 February 1894.

Bristol, 1885 or earlier. Published by Gospel Temperance Union to raise money.

Cambridge, 1890. *Drink Map of Cambridge*, an advertising map for public houses.

Cambridge, 1895. Submitted to use at brewster sessions to show that there were too many red dots; published by Cambridge Temperance United.

Cardiff, 1894. *Drink Map of Cardiff*, presented at a meeting of the Council of the Cardiff Social Reform Union.

Cardiff, 1898. Presented as part of a lecture on social reform.

Chatham, 1894. Used to persuade magistrates to stop granting licences.

Cork, 1891. Creation contemplated because of so many drinking establishments in close proximity to each other; in the notes of an Irish Temperance League meeting.

Coventry, 1885. *Drink Plan of Coventry*.

Derby, 1883. *Map of the Boro of Derby shewing the number and position of Houses Licensed for the Sale of Intoxicating Drinks.*

Derby, 1887. *Map of the Boro of Derby shewing the number and position of Houses Licensed for the Sale of Intoxicating Drinks.*

Dublin, 1891. Based on an offer to make a drink map of Dublin overlaying an Ordnance Survey (refused). There are other mentions of a drink map of Dublin.

Dundee, 1883. Said to show a ratio of one licensed establishment for every 310 inhabitants.

Eastboune, 1859. Parish base map affixed with licensed houses to show magistrates the distances between them.

Edinburgh, 1923. *The Heart of Edinburgh* (showing public houses and licensed grocers).

England, 1878. *Map of Drunkenness.*

England and Wales, 1893. *Workhouse Drink Map of England and Wales* published by the Workhouse Drink Reform League.

England and Wales, 1899. *Geographical Distribution of Drunkenness*, in *The Temperance Problem and Social Reform* by Joseph Rowntree and Arthur Sherwell (1st through 6th edns).

England and Wales, 1900. *Geographical Distribution of Drunkenness*, Plate II, in *The Temperance Problem and Social Reform* by Joseph Rowntree and Arthur Sherwell (7th through 9th edns).

England and Wales, 1904. *Distribution of Crime & Drunkenness in England & Wales 1902* (two maps shown side by side on one sheet).

Evesham, 1894.

Exeter, 1886. *Exeter and Plymouth Gazette* reported on the map, noting 220 licensed houses or one for every 39 families, and 67 within a 300-yard circle of the Cathedral at the centre.

Falkingham, 1887. Introduced by Mr W. Ward of the International Organisation of Good Templars of Leicester at the annual business meeting of the Falkingham Temperance Society.

Glasgow, 1859. Map of Glasgow with public houses, presented at a drinking fountain support meeting.

Glasgow, 1884. *New Plan of Glasgow with suburbs … showing the Distribution of Public Houses Licensed Grocers, Churches and Branches of the 'G.U.Y.M.C.A.'*

Great Yarmouth, 1882. *Map of Great Yarmouth Showing the Position of the Licensed Houses.*

Hastings, 1891. *Drink Map of Hastings.*

Kings Lynn, 1892. *Map of the Borough of King's Lynn Showing The Places Licensed for the Sale of Intoxicating Drinks,* published by the Kings Lynn Vigilance Committee.

Leeds, 1889. *The public-house map of the borough of Leeds,* made by Mr Latchmore of the Vigilante Committee and showing 355 licensed public houses, 422 beer houses and 362 drink shops for a total of 1,139 in black crosses, dots and triangles, or 1 for every 115 adults.

Leicester, 1886. *Drink Map of Leicester,* compiled and issued by the Leicestershire & District Temperance Union.

Lincoln, 1890. An attempt was made to introduce it at a brewster session but the magistrates refused to consult it and then renewed every licence application.

Liverpool, 1874. *The licensed victualler's chart and bona fide traveller's guide,* designed by John McGahey to help drinkers determine if they had travelled far enough to be allowed after-hours intoxicants. A string affixed by wax seal to the map provided a scaled card to help the traveller measure.

Liverpool, 1875. *Map of the Most Unhealthy District in Liverpool.*

Liverpool, 1875. *Comparative Frontages of Public Houses and Private Houses, &c; within 100 Yards of St John's Market.*

Liverpool, 1875. *Map of Public Houses Round the Sailors' Home, Liverpool.*

Liverpool, 1875. *Map shewing Licensed Premises within 200 Yards of Town Hall, Liverpool.*

Liverpool, 1876. *Drink Map of Liverpool,* sent to every member of the House of Commons.

Liverpool, 1883. *Mawdsley's Map of the City of Liverpool and Suburbs, 1883.* Title on verso: *Map of the City of Liverpool, with the Licensed Public-Houses, Beer-Shops, Grocers, Confectioners and other Licenses, marked thereon.* Prepared by Nathaniel Smyth, secretary of the Liverpool Popular Control and Sunday Closing Association (with letterpress).

Liverpool, 1891. *Mawdsley's Map of the City of Liverpool and Suburbs, 1883.* Title on verso: *Map of the City of Liverpool, with the Licensed Public-Houses, Beer-Shops, Grocers, Confectioners and other Licenses, marked thereon.* (Red stamp on upper left corner, 'Map of the City of Liverpool, with the Licensed Public-Houses, Beer-Shops, Grocers, Confectioners, and other Licenses'; Verso blank).

Liverpool, 1899. *'B' Division. Shewing Licensed Houses within the Division.*

Liverpool, 1899. *'C' Division. Shewing Licensed Houses within the Division.*

Liverpool, 1899. *'E' Division. Shewing Licensed Houses within the Division.*

Liverpool, 1899. *'G' Division. Shewing Licensed Houses within the Division.*

Liverpool, 1900. *Drink Map of Liverpool. Thick with Licensed Houses. Their number and Positions.* (Printed in a newspaper.)

London, 1860. *Map Shewing the Number of Public-Houses in the Metropolis* accompanying the pamphlet *Drunkenness, as an indirect cause of crime* by John Taylor.

London, 1860. *Map Shewing the Number of Public-Houses in the Metropolis,* a larger version of the map above, exhibited in the entrance hall of the Metropolitan Free Drinking Fountain Association.

London, 1878. *One Half-mile square in the Heart of London.*

London, 1878. *One Half-mile square in the Heart of London,* a larger version of the map above exhibited as a wall-size backdrop for Dr Nichols's temperance lectures.

London, 1884. *The Modern Plague of London showing the public houses as specified in the London Directory.*

London, 1886. *The Modern Plague of London showing the public houses as specified in the London Directory.*

London, 1887. *The Modern Plague of London showing the public houses as specified in the London Directory.*

London, 1892, map of London produced by Sidney Webb nicknamed 'London's Scarlet Fever'.

London, 1899. *Map No. 1. Soho District, W., Royal Commission on the Liquor Licensing Laws,* Evidence, vol. III, Appendix IV.

London, 1899. *Map No. 2. Whitechapel Road, E., Royal Commission on the Liquor Licensing Laws,* Evidence, vol. III, Appendix IV.

London, 1899. *Map No. 3. Fitzroy Square District, W. Royal Commission on the Liquor Licensing Laws,* Evidence, vol. III, Appendix IV.

London, 1899. *Map No. 4. Soho, St. Martins, St. Giles Etc., Royal Commission on the Liquor Licensing Laws,* Evidence, vol. III, Appendix IV.

London, 1899. *Map No 5. Belsize Park, South Hempstead., Royal Commission on the Liquor Licensing Laws,* Evidence, vol. III, Appendix IV.

London, 1899. *Map No. 6. St. Mary, Somers Town and St. James, Hampstead Road. N.W., Royal Commission on the Liquor Licensing Laws*, Evidence, vol. III, Appendix IV.

London, 1899. *Map No. 7. Pimlico, Belgravia Etc., Royal Commission on the Liquor Licensing Laws*, Evidence, vol. III, Appendix IV.

London, 1899–1900. *Map showing Places of Religious Worship, Public Elementary Schools, and Houses Licensed for the Sale of Intoxicating Drinks.*

London, 1900. *Map Showing Number of Public Houses in a District of Central London* in *The Temperance Problem and Social Reform* by Rowntree and Sherwell, 7th edn.

London, 1900. *The Modern Plague of London showing the public houses as specified in the London Directory.*

Manchester, 1888. *Drink Map of Manchester*, provided to politicians.

Manchester, 1889. *Drink Map of Manchester*, printed in the *Manchester Guardian.*

Manchester, 1889. *Drink Map of Manchester*, red on red with history and purpose in letterpress to the right of the map on its face.

Manchester, 1889. *Map of Manchester indicating the licensed victuallers, beer, wine and other licensed houses.*

Maryport, 1899. Referred to as the 'Drunk Map of Maryport'.

Newcastle, 1882. *Drink Map of Newcastle*, possible first issue intended to be framed and hung in public places.

Newcastle, 1883. *Drink Map of Newcastle*, published by the Newcastle Auxiliary of the United Kingdom Alliance.

Newton Heath, 1840. *Map of Old Licenses at Newton Heath.*

Norwich, 1878. *Drink Map of Norwich*, published by the Norwich Auxiliary of the United Kingdom Alliance.

Norwich, 1886. *Map Showing Public Houses in Norwich.*

Norwich, 1892. *Drink Map of Norwich*, published by the Norfolk & Norwich Gospel Temperance Union.

Norwich, 1903. *Drink Map of Norwich*, published jointly by the Temperance Divisional Council, Norfolk & Norwich Gospel Temperance Union and the Joint Brewster Sessions Committee of the Church of England Temperance Society.

Over Darwen, 1882. No title; provided digitally by Roy Taylor of Darwen. (Location of original unknown.)

Oxford, 1883. *Drink Map of Oxford*, published by the Band of Hope.

Oxford, 1883. *Drink Map of Oxford*, published by the Band of Hope (alternate typeface title).

Oxford, 1883. *Drink Map of Oxford*, framed and glazed version presented at a brewster session.

Preston, 1886. Produced by the Preston and District Band of Hope Union showing 489 licensed premises.

Plymouth, 1893.

St Albans, 1891. Presented at a temperance garden party.

St Leonards, 1891.

Salford, 1892.

Salisbury, 1886. Published by the Salisbury Temperance Association.

Sheffield, 1884. *Drink Map of the Town of Sheffield.*

Sheffield, 1903. *Drink Map of the Town of Sheffield*, 6 foot by 12 foot, framed and at one time located in the Council Chambers.

Southampton, 1878. *Drink Map of Southampton.*

Stafford, 1880. *Drink Map of Stafford.*

Stafford 1883. *Drink Map of Stafford*, published by W.C. Amery; letterpress noted about liquor laws and traffic.

Stockport, 1880. *Drink Map of Stockport.* Published by the Gospel Temperance Union.

Sunderland, 1899. *Drink Map of Sunderland*, used as preface to a United Temperance Society report.

Thetford, 1890. Showing 31 public houses, or one for every 148 inhabitants.

Todmorden, 1890. Prepared by Mr Byles and appearing in the *Bradford Observer.*

Woolwich, 1893.

Yarmouth, 1903. Prepared at the request of licensing justices.

York, 1882. *Drink Map of York* published by the Alliance York Auxiliary.

York, 1901, *Map of York showing position of Licenced Houses*, accompanying *Poverty: A Study of Town Life* by Rowntree.

UNITED STATES

Bangor, Maine, 1899. *Map showing the number and location of Liquor Saloons, etc., in Bangor, Maine.*

Boston, 1899. *Map showing the position of 'licence' and 'no licence' towns and cities in the vicinity of Boston, Massachusetts.*

Chicago, 1893. *If Christ Came to Chicago* (Nineteenth Precinct, First Ward, showing brothels, pawn brokers, saloons and lodging houses).

Lewiston, Maine, 1899. *Map showing the number and location of Liquor Saloons, etc., in Lewiston, Maine.*

Manchester, New Hampshire, 1899. *Map showing the number and location of Liquor Saloons, etc., in Manchester, New Hampshire.*

Nashua, New Hampshire, 1899. *Map showing the number and location of Liquor Saloons, etc., in Nashua, New Hampshire.*

New York, 1883. *Liquordom in New York City (a), (b), (c), (d)*; four maps accompanying a pamphlet of the same title by Robert Graham.

New York, 1886. *Map of New York City*, inside *The Temperance Movement or the Conflict between Man & Alcohol* by Henry William Blair.

Philadelphia, 1901. [Untitled] Lighthouse saloon map of Philadelphia.

Portland, Maine, 1899. *Map showing the number and location of Liquor Saloons, etc., in Portland, Maine.*

United States, 1899. *Map showing the extent to which the right of local option exists in the United States and the way in which it can be exercised.*

Waterville, Maine, 1899. *Map showing the number and location of Liquor Saloons, etc., in Waterville, Maine.*

NOTES

ONE

1. Thomas Low Nichols, *One Half-mile Square in the Heart of London,. (A [temperance] lecture delivered in Salisbury Hall.)*, British Library, Historical Print Editions, London, 1878, p. 12.
2. Ibid., p. 13.

TWO

1. Joseph Livesey, *The Staunch Teetotaler,* Tweedie, London, 1869, p. 308.
2. Sidney Smith in a contemporary letter quoted in *The Pub and the People*, A Worktown Study by Mass Observation, Victor Gallancz, London, 1943, p. 84.
3. Sir Wilfrid Lawson, *Wit and wisdom of Sir Wilfrid Lawson: being selections from his speeches, 1865–1885*, Simpkin, Marshall, London, 1886, p. 49.
4. Sir Wilfrid Lawson, *A Memoir*, ed. George W.E. Russell, Smith, Elder, London, 1909, p. 59.
5. James Nicholls, *The Politics of Alcohol*, 2009, p. 84.
6. Sir Wilfrid Lawson, *A Memoir*, ed. George W.E. Russell, Smith, Elder, London, 1909, p. 61.
7. Charles Buxton, *North British Review* 22, February 1855, p. 11.
8. *Liverpool Mail,* 2 May 1874, quoting the Licensing Act of 1872.

THREE

1. Frederick G. Hindle, *The Legal Status of Licensed Victuallers*, 4th edn, Stevens & Sons, London, 1884, p. 34.
2. Ibid., p. 28.

SIX

1. *Royal Commission on Liquor Licensing Laws, Final Report: Minutes of Evidence,* vol. III p. 172.
2. *Charles Booth, Life and Labour of the People of London, Final Volume,* p. 79.
3. *Royal Commission on Liquor Licensing Laws, Final Report: Minutes of Evidence,* vol. III p. 173.
4. *Charles Booth, Life and Labour of the People of London, Final Volume,* p. 69.
5. See L. Vaughan, *Mapping Society: The Spatial Dimensions of Social Cartography,* ch. 4, UCL Press, 2018.
6. *Royal Commission on Liquor Licensing Laws, Final Report: Minutes of Evidence,* vol. III p. 168.
7. Ibid., p. 167.

FURTHER READING

Beckingham, D., *The Licensed City: Regulating Drink in Liverpool, 1830–1920*, Liverpool University Press, Liverpool, 2017.

Black, R., A *Talent for Humanity: The Life and Work of Lady Henry Somerset*, Anthony Rowe, Chippenham, 2010.

Briggs, A., *Victorian Cities: A Brilliant and Absorbing History of Their Development*, Penguin, London, 1990.

Brown, P., *The Pub: A Cultural Institution – from Country Inns to Craft Beer Bars and Corner Locals*, Jacqui Small, 2016.

Buxton, C., *North British Review* 22, February 1855.

Buxton, S., ed., *The Imperial Parliament Series: Local Option*, Swan Sonnenschein, Le Bas & Lowrey, 1885.

Cooke, A., *A History of Drinking: The Scottish Pub since 1700*, Edinburgh University Press, Edinburgh, 2015.

Cornell, M., *Beer. The Story of the Pint*, Headline, London, 2003.

Cornell, M., *Amber Gold & Black: The History of Britain's Great Beers*, History Press, Cheltenham, 2010.

Dingle, A.E., *The Campaign for Prohibition in Victorian England: The United Kingdom Alliance 1872–1895*, Croom Helm, London, 1980.

Fleet, C., and D. MacCannell, *Edinburgh: Mapping the City*, Birlinn, Edinburgh, 2014.

Gutzke, D., *Protecting the Pub: Brewers and Publicans against Temperance*, Boydell Press, Suffolk, 1989.

Hackwood, F., *Inns, Ales and Drinking Customs of Old England*, Bracken Books, London, 1985.

Hamer, D.A., *The Politics of Electoral Pressure: A Study in the History of Victorian Reform Agitations*, Humanities Press, Atlantic Highlands NJ, 1977.

Harrison, B., *Drink & the Victorian: The Temperance Question in England 1815–1872*, 2nd edn, Keele University Press, 1994.

Hayler, M., *The Vision of a Century: The United Kingdom Alliance in Historical Retrospect*, United Kingdom Alliance, London, 1953.

Hindle, F.G., *The Legal Status of Licensed Victuallers*, 4th edn, Stevens & Sons, London, 1884.

Kain, R., and R. Oliver, *British Town Maps: A History*, British Library, London, 2015.

Kneale, J., 'Good, Homely, Troublesome or Improving? Historical Geographies of Drinking Places, *c.*1850–1950', *Geography Compass*, vol. 15, no. 3, 2021.

Kneale, J., and S. French, 'Mapping Alcohol: Health, Policy and the Geographies of Problem Drinking in Britain', *Drugs: Education, Prevention and Policy*, vol. 15, no. 3, 2008, 233–49.

Livesey, J., *The Staunch Teetotaler*, Tweedie, London, 1869.

MacCannell, D., *Oxford: Mapping the City*, Birlinn, Edinburgh, 2016.

McAllister, A., 'The Alternative World of the Proud Non-Drinker: Nineteenth-century Public Displays of Temperance', *The Social History of Alcohol and Drugs*, vol. 28, no. 2, 2014, 161–79.

Moore, J., Glasgow: *Mapping the City*, Berlinn, Edinburgh, 2015.

Nicholls, J., *The Politics of Alcohol: A History of the Drink Question in England*, Manchester University Press, Manchester, 2009.

Nichols, T. *Forty Years of American Life 1821–1861*, Stackpole Sons, New York, 1874.

Nichols, T., *One Half-mile Square in the Heart of London. (A [temperance] lecture delivered in Salisbury Hall.)*, British Library, Historical Print Editions, London, 1878.

Niessen, O.C., *Aristocracy, Temperance and Social Reform: The Life of Lady Henry Somerset*, I.B. Tauris, London, 2007.

Pattinson, R., *Bitter!*, Kilderkin, Amsterdam, 2013.

Pattinson, R., *The Home Brewer's Guide to Vintage Beer: Rediscovered Recipes for Classic Brews dating from 1800 to 1965*, Quarry Books, Beverly MA, 2014.

Roberts, J.S., *Drink, Temperance and the Working Class in Nineteenth-Century Germany*, Allen & Unwin, London, 1984.

Rowntree, B.S., *Poverty: A Study of Town Life*, Macmillan, London, 1901.

Rowntree, J., and A. Sherwell, *The Temperance Problem and Social Reform*, 2nd edn, Hodder & Stoughton, London, 1899, and 7th edn, Truslove Hanson & Comba, New York, 1900.

Royal Commission on Liquor Licensing Laws, Final Report: Minutes of Evidence, vol. III, 1899.

Russell, G., ed., *Sir Wilfrid Lawson: A Memoir*, Smith, Elder, London, 1909.

Shaw, J.G., *History and Traditions of Darwen and its People*, J. and G. Toulmin, Printers, Blackburn, 1889.

Spence, F.S., *The Facts of the Case: a summary of the most important evidence and argument presented in the report of the Royal Commission on the Liquor Traffic*, Newton & Treloar, Toronto, 1896.

Spiller, B., *Victorian Public Houses*, David & Charles, Newton Abbot, 1972.

Vaughan, L., *Mapping Society: The Spatial Dimensions of Social Cartography*, UCL Press, London, 2018.

Wyke, T., B. Robson and M. Dodge, *Manchester: Mapping the City*, Birlinn, Edinburgh, 2018.

Charles Booth's notebooks are available online at the London School of Economics, while Lady Somerset's diaries have not been found. Many of her letters to her family and to notable reformers are helpful in piecing together the impact of her work in addition to newspaper articles in the UK and abroad.

The British Newspaper Archive has been a rich resource of history about drink maps, brewster sessions, the Victorian era and some very strange stories.

PICTURE CREDITS

10 (16); (p. 90) Bodleian Library, John Johnson Collection, Wines and Spirits 1 (20)

21 Sheffield Libraries and Archives/© PictureSheffield.com.

22 Sheffield Libraries and Archives/© PictureSheffield.com

23 Reproduced with the permission of the Library of Birmingham, MAP/149421

24 The Record Office for Leicestershire, Leicester & Rutland, L912 Box D

25 Bodleian Library, John Johnson Collection, Temperance folder 1, Temperance organisations, provincial.

26 Bodleian Library, John Johnson Collection, Temperance folder 1, Temperance organisations, provincial.

27 Copyright of the University of Manchester, Map Collection (Flat) C17:70 Manchester (3)

28 Bodleian Library, Per. 16871 c.2, p. 464

29 Bodleian Library, Johnson b.108 (2), Plate 1

30 Bodleian Library, Johnson b.108 (2), Plate 2

31 Chetham's Library, Manchester

32 Image courtesy of Norfolk County Council Library and Information, XII 1878 Ch, p. 233

33 Author's collection

34 Bodleian Library, John Johnson Collection, Temperance Boxes 2 and 6

35 Author's collection, Royal Commission on Licensing Laws Evidence Appendix IV

36 Author's collection, Royal Commission on Licensing Laws Evidence Appendix IV

37 David Rumsey Map Collection, David Rumsey Map Center, Stanford Libraries

38 Museum of London, 64.96

39 Liverpool Central Library, (Planfile Map Case 40), f912.1889

40 Reproduced with the permission of the National Library of Scotland, Acc.10222/PR/65a, Folio 50

INDEX

Entries in *italics* refer to illustrations